Student's Guide to
Calculus
by J. Marsden and A. Weinstein
Volume I

Frederick H. Soon

Student's Guide to
Calculus
by J. Marsden and A. Weinstein
Volume I

With 172 Illustrations

Springer-Verlag
New York Berlin Heidelberg Tokyo

AMS Subject Classification: 26-01

Library of Congress Cataloging-in-Publication Data
Soon, Frederick H.
   Student's Guide to Calculus, volume 1
   1. Calculus.   I. Marsden, Jerrold E.   Calculus I.
II. Title.
QA303.S774   1985   515   85-17198

9 8 7 6 5 4 3 2 1

ISBN-13: 978-0-387-96207-8   e-ISBN-13: 978-1-4612-5146-0
DOI: 10.1007/978-1-4612-5146-0

Dedicated to:

Henry, Ora, Dennis, and Debbie

This Student Guide is exceptional, maybe even unique, among such guides in that its author, Fred Soon, was actually a student user of the textbook during one of the years we were writing and debugging the book. (He was one of the best students that year, by the way.) Because of his background, Fred has taken, in the Guide, the point of view of an experienced student tutor helping you to learn calculus. While we do not always think Fred's jokes are as funny as he does, we appreciate his enthusiasm and his desire to enter into communication with his readers; since we nearly always agree with the mathematical judgements he has made in explaining the material, we believe that this Guide can serve you as a valuable supplement to our text.

To get maximum benefit from this Guide, you should begin by spending a few moments to acquaint yourself with its structure. Once you get started in the course, take advantage of the many opportunities which the text and Student Guide together provide for learning calculus in the only way that any mathematical subject can truly be mastered — through attempting to solve problems on your own. As you read the text, try doing each example and exercise yourself before reading the solution; do the same with the quiz problems provided by Fred.

Fred Soon knows our textbook better than anyone with the (possible) exception of ourselves, having spent hundreds of hours over the past ten years assisting us with its creation and proofreading. We have enjoyed our association with him over this period, and we hope now that you, too, will benefit from his efforts.

Jerry Marsden

Alan Weinstein

HOW TO USE THIS BOOK

As the title implies, this book is intended to guide the student's study
of calculus. Realizing that calculus is not the only class on the college
student's curriculum, my objective in writing this book is to maximize understanding with a minimum of time and effort.

For each new section of the text, this student guide contains sections
entitled Prerequisites, Prerequisite Quiz, Goals, Study Hints, Solutions to
Every Other Odd Exercise, Section Quiz, Answers to Prerequisite Quiz, and
Answers to Section Quiz. For each review section, I have included the solutions to every other odd exercise and a chapter test with solutions.

A list of prerequisites, if any, is followed by a short quiz to help
you decide if you're ready to continue. If some prerequisite seems vague to
you, review material can be found in the section or chapter of the text listed
after each prerequisite. If you have any difficulty with the simple prerequisite quizzes, you may wish to review.

As you study, keep the goals in mind. They may be used as guidelines
and should help you to grasp the most important points.

The study hints are provided to help you use your time efficiently.
Comments have been offered to topics in the order in which they appear in the
text. I have tried to point out what is worth memorizing and what isn't. If
time permits, it is advisable to learn the derivations of formulas rather than
just memorizing them. You will find that the course will be more meaningful

to you and that critical parts of a formula can be recalled even under the stress of an exam. Other aspects of the study hints include clarification of text material and "tricks" which will aid you in solving the exercises. Finally, please be aware that your instructor may choose to emphasize topics which I have considered less important.

Detailed solutions to every other odd exercise, i.e., 1,5,9, etc. are provided as a study aid. Some students may find it profitable to try the exercises first and then compare the method employed in this book. Since the authors of the text wrote most of the exercises in pairs, the answers in this book may also be used as a guide to solving the corresponding even exercises. In order to save space, fractions have been written on one line, so be careful about your interpretations. Thus, $1/x + y$ means y plus $1/x$, wherea $1/(x + y)$ means the reciprocal of $x + y$. Transcendental functions such as cos, sin, ln, etc. take precedence over division, so cos ax/b means take the cosine of ax and then divide by b, whereas cos (ax/b) has an unambiguou meaning. ln $a/2$ means half of ln a, not the natural logarithm of $a/2$. Also, everything in the term after the slash is in the denominator, so $1/2\int xdx + 1$ means add 1 to the reciprocal of $2\int xdx$. It does not mean add 1 to half of the integral. The latter would be denoted $(1/2)\int xdx + 1$.

Section quizzes are included for you to evaluate your mastery of the material. Some of the questions are intended to be tricky, so do not be discouraged if you miss a few of them. The answers to these "hard" questions should add to your knowledge and prepare you for your exams. Since most students seei to fear word problems, each quiz contains at least one word problem to help you gain familiarity with this type of question.

Finally, answers have been provided to both the prerequisite and section quizzes. If you don't understand how to arrive at any of the answers, be sure to ask your instructor.

In the review sections, I have written more questions and answers which may appear on a typical test. These may be used along with the section quizzes to help you study for your tests.

Since Calculus was intended for a three semester course, I have also included three-hour comprehensive exams at the end of Chapters 3, 6, 9, 12, 15, and 18. These should help you prepare for your midterms and final examinations. Best of luck with all of your studies.

ACKNOWLEDGEMENTS

Several individuals need to be thanked for helping to produce this book. I am most grateful to Jerrold Marsden and Alan Weinstein for providing the first edition of Calculus from which I, as a student, learned about derivatives and integrals. Also, I am deeply appreciative for their advice and expertise which they offered during the prepartion of this book. Invaluable aid and knowledgeable reviewing were provided by my primary assistants: Stephen Hook, Frederick Daniels, and Karen Pao. Teresa Ling should be recognized for laying the groundwork with the first edition of the student guide. Finally, my gratitude goes to my father, Henry, who did the artwork; to Charles Olver and Betty Hsi, my proofreaders; and to Ruth Edmonds, whose typing made this publication a reality.

Frederick H. Soon
Berkeley, California

CONTENTS

CHAPTER R

REVIEW OF FUNDAMENTALS

R.1 Basic Algebra: Real Numbers and Inequalities

PREREQUISITES

1. There are no prerequisites for this section other than some high school
    algebra and geometry; however, if the material presented in this section
    is new to you, it would be a good idea to enroll in a precalculus course.
    This section is intended to be a review.

PREREQUISITE QUIZ

1. Orientation quizzes A and B in the text will help you evaluate your
    preparation for this section and this course.

GOALS

1. Be able to factor and expand common mathematical expressions.

2. Be able to complete a square.

3. Be able to use the quadratic formula.

4. Be able to solve equations and inequalities.

STUDY HINTS

1. Common identities. Know how to factor $a^2 - b^2$. It is a good idea to
    memorize the expansion of $(a + b)^2$ and $(a + b)^3$. Note that $(a - b)^2$
    can be obtained by substituting $-b$ for b . $(a - b)^3$ can be similarly
    expanded. These identities are useful for computing limits in Section 1.2
    and Chapter 11 .

2. Factoring. This is a technique that is learned best through practice. A
    good starting point is to find all integer factors of the last term (the
    constant term). Once you find a factor for the original polynomial, use
    long division to find a simpler polynomial to factor. This will be impor-
    tant for partial fractions in Chapter 10 and for computing limits.

3. Completing the square. Don't memorize the formula. Practice until you
    learn the technique. Note that adding $(b/2a)^2$ to $x^2 + bx/a$ forms a
    perfect square. This technique will be very important for integration
    techniques introduced in Chapter 10.

4. Quadratic formula. · It is recommended that you memorize this formula. It
    is used in many applications in various disciplines such as engineering,
    economics, medicine, etc. This formula may also be used to solve equa-
    tions of the form $Ax^4 + Bx^2 + C = 0$ by solving for $y = x^2$ and taking
    square roots to get x .

5. Square roots. Note that, unless otherwise stated, square roots are under-
    stood to be nonnegative. $\sqrt{0}$ is equal to zero.

6. Inequalities. It is essential to have a good handle on manipulating
    inequalities. Without this, you will not have a good understanding of
    some of the basic theory of calculus. Don't forget to reverse the direc-
    tion of the inequality sign when you multiply by a negative number.

SOLUTIONS TO EVERY OTHER ODD EXERCISE

1. $8/6 - 9/4 = -11/12$ is a rational number. Since the denominator cannot be reduced to one, it is neither a natural number nor an integer.

5. $(a - 3)(b + c) - (ac + 2b) = (ab - 3b + ac - 3c) - (ac + 2b) = ab - 5b - 3c$.

9.   We can use Example 2 with b replaced by $-b$ to get $a^3 + 3a^2(-b) +$ $3a(-b)^2 + (-b)^3 = a^3 - 3a^2b + 3ab^2 - b^3$. Alternatively, write $(a - b)^3 =$ $(a - b)^2(a - b) = (a^2 - 2ab + b^2)(a - b) = a^3 - 3a^2b + 3ab^2 - b^3$.

13. We know that $(x + a)(x + b) = x^2 + (a + b)x + ab$. The factors of 6 are ± 1 , ± 2 , ± 3 , and ± 6 . By choosing $a = 2$ and $b = 3$, we get $a + b = 5$. Thus, $x^2 + 5x + 6 = (x + 2)(x + 3)$.

17. First we factor out 3 to get $3(x^2 - 2x - 8)$. We know that the factors of -8 are ± 1 , ± 2 , ± 4 , and ± 8 . As in Exercise 13, we look for a and b so that $(a + b)x$ is the middle term. In this case, $a = -4$ and $b = 2$. Thus, $3x^2 - 6x - 24 = 3(x - 4)(x + 2)$.

21. $2(3x - 7) - (4x - 10) = 0$ simplifies to $(6x - 14) - (4x - 10) =$ $2x - 4 = 0$, i.e., $2x = 4$. Dividing by 2 yields $x = 2$.

25. The right-hand side is $(x - 1)(x^2 + x + 1) = x(x^2 + x + 1) - 1(x^2 + x + 1) =$ $x^3 + x^2 + x - x^2 - x - 1 = x^3 - 1$, which is the left-hand side.

29. (a) By factoring, we get $x^2 + 5x + 4 = (x + 4)(x + 1) = 0$. If any factor equals zero, the equation is solved. Thus, $x = -4$ or $x = -1$.

     (b) By using the method of completing the square, we get $0 = x^2 + 5x + 4 =$ $(x^2 + 5x + 25/4) + (4 - 25/4) = (x + 5/2)^2 - 9/4$. Rearrangement yields $(x + 5/2)^2 = 9/4$, and taking square roots gives $x + 5/2 =$ $\pm 3/2$. Again, $x = -4$ or -1 .

     (c) $a = 1$, $b = 5$, and $c = 4$, so the quadratic formula gives $x = (-5 \pm \sqrt{25 - 4(1)(4)})/2(1) = (-5 \pm 3)/2 = -4$ or -1 .

33. We use the quadratic formula with $a = -1$, $b = 5$, and $c = 0.3$.
This gives $x = (-5 \pm \sqrt{25 + 1.2})/(-2) = (5 \pm \sqrt{26.2})/2$. These are the
two solutions for x .

37. $x^2 + 4 = 3x^2 - x$ is equivalent to $0 = 2x^2 - x - 4$. Using the quadrati
formula with $a = 2$, $b = -1$, and $c = -4$, we get $x = (1 \pm \sqrt{1 + 32})/$
$(1 \pm \sqrt{33})/4$.

41. We apply the quadratic formula with $a = 2$, $b = 2\sqrt{7}$, and $c = 7/2$ to
get $x = (2\sqrt{7} \pm \sqrt{28 - 28})/4 = \sqrt{7}/2$. Thus, the only solution is $x = \sqrt{7}/$

45. We add b to both sides to obtain $a + c > 2c$. Then we subtract c to
get $a > c$.

49. $b(b + 2) > (b + 1)(b + 2)$ is equivalent to $b^2 + 2b > b^2 + 3b + 2$.
Subtracting $b^2 + 2b$ from both sides leaves $0 > b + 2$. Subtracting 2
yields $-2 > b$.

53. (a) Dividing through by a in the general quadratic equation yields
$x^2 + (b/a)x + c/a = 0$. Add and subtract $(b/2a)^2$ to get
$(x^2 + (b/a)x + b^2/4a^2) + (c/a - b^2/4a^2) = 0 = (x + b/2a)^2 +$
$(4ac/4a^2 - b^2/4a^2)$, i.e., $(x + b/2a)^2 = (b^2 - 4ac)/4a^2$. Taking
square roots gives $x + b/2a = \pm\sqrt{b^2 - 4ac}/2a$, and finally
$x = (-b \pm \sqrt{b^2 - 4ac})/2a$.

(b) From the quadratic formula, we see that there are no solutions if
$b^2 - 4ac < 0$. If $b^2 - 4ac > 0$, there are two distinct roots.
However, if $b^2 - 4ac = 0$, there are two roots, which both equal
$-b/2a$. This only occurs if $b^2 = 4ac$.

SECTION QUIZ

1.   Factor: (a) $x^4 + 2x^2 + 1$

            (b) $2x^4 - x^2 - 1$

            (c) $x^6 - 1$

2.   Apply the expansion of $(a + b)^3$ to expand $(3x - 2)^3$.

3.   Use the quadratic formula to solve $x^5 + 3x^3 - 5x = 0$.

4.   Sketch the solution of   (a) $x^2 + 3x + 2 < 0$

                          (b) $x^2 + 3x + 2 \geq 0$

5.   Find the solution set of $x^3 \geq x$.

6.   The first King of the Royal Land of Mathematica has decreed that the
     first young lady to answer the following puzzle shall rule at his side.
     The puzzle is to compute the product of all solutions to
     $x^3 + 2x^2 - x - 2 = 0$. Then, divide by the length of the finite interval
     for which $x^3 + 2x^2 - x - 2 \geq 0$. What answer would make a lady Queen
     of Mathematica?*

ANSWERS TO SECTION QUIZ

1.   (a) $(x^2 + 1)^2$

     (b) $(2x^2 + 1)(x + 1)(x - 1)$

     (c) $(x^3 + 1)(x^2 + x + 1)(x - 1)$

2. $27x^3 - 54x^2 + 36x - 8$

*Dear Reader: I realize that many of you hate math but are forced to com-
plete this course for graduation. Thus, I have attempted to maintain interest with "entertaining" word problems. They are not meant to be insulting to your intelligence. Obviously, most of the situations will never
happen; however, calculus has several practical uses and such examples are
found throughout Marsden and Weinstein's text. I would appreciate your
comments on whether my "unusual" word problems should be kept for the next
edition.

3. 0 , $\pm[(-3 + \sqrt{29})/2]^{1/2}$

4.   (a)

$\begin{array}{cc} -2 & -1 \end{array}$ x

(b)

$\begin{array}{cc} -2 & -1 \end{array}$ x

5. $-1 \leq x \leq 0$ and $x \geq 1$

6.   2

R.2   Intervals and Absolute Values

PREREQUISITES

1.   Recall how to solve inequalities.   (Section R.1)

PREREQUISITE QUIZ

1.   Solve the following inequalities:

   (a) $-x < 1$

   (b) $5x + 2 \geqslant x - 3$

GOALS

1.   Be able to express intervals using symbols.

2.   Be able to manipulate absolute values in equations and inequalities.

STUDY HINTS

1.   Notation. A black dot means that the endpoint is included in the
     interval. In symbols, a square bracket like this "[" or like this "]"
     is used. A white circle means that the endpoint is not included in the
     interval; it is represented by a parenthesis like this "(" or this ")".

2.   More notation. In the solution to Example 3, some students will get lazy
     and write the solution as $-1 > x > 3$. This gives the implication that
     $-1 > 3$, which is false.

3.   Inequalities involving absolute values. Study Example 5 carefully.
     Such inequalities are often used in Chapter 11. Note that $|x - 8| \leq 3$
     is the same as $-3 \leq x - 8 \leq 3$.

4.   Triangle inequality. $|x + y| \leq |x| + |y|$ will be useful in proving limit
     theorems in Chapter 11. The name is derived from the fact that two sides
     of a triangle are always longer than the third side.

SOLUTIONS TO EVERY OTHER ODD EXERCISE

1. (a) This is true because $-8 \leqslant -7 \leqslant 1$.

    (b) This is false because $5 < 11/2$.

    (c) This is true because $-4 < 4 \leqslant 6$.

    (d) This is false because $4 < 4 < 6$ is false. 4 is not less than 4 .

    (e) This is true because $3/2 + 7/2 = 5$, which is an integer.

5. $x + 4 \geqslant 7$ implies $x \geqslant 3$. In terms of intervals, $x \in [\, 3, \infty\,)$.

9. $x^2 + 2x - 3 > 0$ implies $(x + 3)(x - 1) > 0$. In one case, we need
    $x > -3$ and $x > 1$, i.e., $x > 1$. On the other hand, we can also
    have $x < -3$ and $x < 1$, i.e., $x < -3$. In terms of intervals,
    $x \in (-\infty, -3)$ or $x \in (1, \infty)$.

13. The absolute value is the distance from the origin. $|3 - 5| = |-2|$.
     Since $-2 < 0$, we change the sign to get $|-2| = 2$.

17. The absolute value of $|x| = x$ if $x > 0$. Thus, $|3 \cdot 5| = |15| = 15$.

21. Using the idea of Example 4(d), the only two solutions are $x = \pm 8$.

25. $x^2 - x - 2 > 0$ implies $(x - 2)(x + 1) > 0$. In one case, we have
     $x > 2$ and $x > -1$, i.e., $x > 2$. On the other hand, we can have
     $x < 2$ and $x < -1$, i.e., $x < -1$. The solution is all x except
     for x in $[-1, 2]$. Using the idea of Example 6, we want to eliminate
     $|x - 1/2| \leqslant 3/2$. Therefore, the solution is $|x - 1/2| > 3/2$.

29. $|x| < 5$ implies $-5 < x < 5$, so x belongs to the interval $(-5, 5)$.

33. The midpoint of $(-3, 3)$ is 0 and the length of half of the interval
     is 3 . Therefore, by the method of Example 6, we get $|x - 0| = |x| < 3$

37. The midpoint of $[-8, 12]$ is 2 and the length of half of the interval
     is 10 . Therefore, by the method of Example 6, we get $|x - 2| \leqslant 10$.
     Equality is allowed as a possibility since the endpoints are included.

41. If $x \geq 0$, then $x^3 \geq 0$. In this case, the cube root of x^3 is still
     positive and equals x . If $x < 0$, then $x^3 < 0$. In this case, the
     cube root of x^3 is negative and equals x . Thus, independent of the
     sign of x , we have $x = \sqrt[3]{x^3}$.

SECTION QUIZ

1. Express the possible solutions of $x^3 - x \geq 0$ in terms of intervals.

2. Which of the following is true?

    (a) $\sqrt{-1} = -1$.   (b) $\sqrt{0} = 0$.

    (c) $\sqrt{x^2} = x$ if $x < 0$.   (d) $\sqrt{x^2} = |x|$ for all $x \neq 0$.

3. Solve $|x - 5| \geq 5$.

4. The school bully has selected you to help him do his homework problem,
    which is to solve $x^2 - 6x + 5 \geq 0$. You determine that the solution
    is $x \leq 1$ or $x \geq 5$. The bully understands, but then you turn the
    tables on him and tell him, "Look, $1 \geq x$ and $x \geq 5$, so $1 \geq x \geq 5$."
    He turns in this answer.

    (a) Explain why $1 \geq x \geq 5$ is incorrect.

    (b) Write the correct answer in terms of intervals.

    (c) Write your answer in the form $|x - a| \geq b$ for constants a and b .

5. An architect is building an arched doorway whose height at x feet from the
    left base is $x - 0.1x^2$.

    (a) How wide is the doorway at position $x = 1.5$?

    (b) Express the width at $x = 1.5$ in terms of absolute values.

ANSWERS TO PREREQUISITE QUIZ

1.   (a) $x > -1$

     (b) $x \geqslant -5/4$

ANSWERS TO SECTION QUIZ

1. $[-1,0]$ and $[1,\infty)$

2.   b   and   d

3. $x \leqslant 0$ and $x \geqslant 10$

4.   (a)   The implication is that $5 \leqslant 1$.

     (b) $(-\infty, 1]$ and $[5, \infty)$

     (c) $|x - 3| \geqslant 2$

5.   (a)   7

     (b) $|x - 5| \leqslant 7/2$

R.3 Laws of Exponents

PREREQUISITIES

1. There are no prerequisites for this section beyond some high school algebra; however, if the material presented in this section is new to you, it would be a good idea to enroll in a precalculus course. This section is intended to be a review.

GOALS

1. Be able to simplify expressions involving exponents.

STUDY HINTS

1. Integer exponents. The laws of exponents should be memorized. The "common sense" method preceding Example 1 serves as a useful check if you are unsure of your answer; however, it can slow you down during an exam.

2. Definitions. Know that $b^{-n} = 1/b^n$, $b^0 = 1$, and $b^{1/n} = \sqrt[n]{b}$.

3. Rational exponents. These laws are the same as those for integer powers.

4. Rationalization. Example 6 demonstrates a technique which is useful in the study of limits. The idea is to use the fact that $(a + b)(a - b) = a^2 - b^2$ to eliminate radicals from the denominator.

SOLUTIONS TO EVERY OTHER ODD EXERCISE

1. By the law $(bc)^n = b^n c^n$, we have $3^2 (1/3)^2 = (3 \cdot 1/3)^2 = (1)^2 = 1$.

5. $[(4 \cdot 3)^{-6} \cdot 8]/9^3 = 8/4^6 \cdot 3^6 \cdot (3^2)^3 = 2^3/(2^2)^6 \cdot 3^6 \cdot 3^6 = 2^3/2^{12} \cdot 3^{12} = 1/2^9 \cdot 3^{12}$.

9. $9^{1/2} = \sqrt{9} = 3$.

13. $2^{5/3}/4^{7/3} = 2^{5/3}/(2^2)^{7/3} = 2^{5/3}/2^{14/3} = 2^{5/3 - 14/3} = 2^{-3} = 1/2^3 = 1/8$.

17. Apply to distributive law to get $(x^{3/2} + x^{5/2})x^{-3/2} = (x^{3/2})(x^{-3/2}) +$
    $(x^{5/2})(x^{-3/2}) = x^{3/2 - 3/2} + x^{5/2 - 3/2} = x^0 + x^1 = 1 + x$.

21. By using the laws of rational exponents, we get $\sqrt[a]{\sqrt[b]{x}} = (\sqrt[b]{x})^{1/a} =$
    $(x^{1/b})^{1/a} = x^{1/ab} = \sqrt[ab]{x}$.

25. Since the price doubles in 10 years, it will double again after 20
    years and the factor is $2 \cdot 2 = 4$. In thirty years, the factor is
    $2 \cdot 2 \cdot 2 = 8$, and in 50 years, the factor of increase is $2 \cdot 2 \cdot 2 \cdot 2 \cdot 2 = 32$

29. The first term factors into $\sqrt{x} \cdot \sqrt{x}$ and $(\sqrt{x} + a)(\sqrt{x} + b) = x^2 +$
    $(a + b)\sqrt{x} + ab$. The last term has factors ± 1 , ± 2 , ± 4 , and ± 8 .
    We want factors a and b such that $ab = -8$ and $a + b = -2$. Thus
    $a = -4$ and $b = 2$. Therefore, the solution is $(\sqrt{x} - 4)(\sqrt{x} + 2) =$
    $(x^{1/2} - 4)(x^{1/2} + 2)$.

33. We will use the corresponding rule for integer powers several times.
    Let $p = m/n$ and $q = k/l$, so $b^{pq} = b^{(m/n)(k/l)} = b^{mk/nl} = \sqrt[nl]{b^{mq}}$.
    Thus, rasing the equation to the nl power gives $(b^{pq})^{nl} = b^{mk}$.
    Now, $(b^p)^q = (b^{m/n})^{k/l} = \sqrt[l]{(\sqrt[n]{b^m})^k}$, so $((b^p)^q)^{nl} = (\sqrt[l]{(\sqrt[n]{b^m})^k})^{ln} =$
    $((\sqrt[n]{b^m})^k)^n = (\sqrt[n]{b^m})^{nk} = ((\sqrt[n]{b^m})^n)^k = (b^m)^k = b^{mk}$. Therefore, $(b^{pq})^{nl} =$
    $((b^p)^q)^{nl}$, and taking the nl^{th} root of both sides gives Rule 2.

    For Rule 3, let $p = m/n$. Then $(bc)^p = (bc)^{m/n} = \sqrt[n]{(bc)^m} =$
    $\sqrt[n]{b^m c^m}$, so $((bc)^p)^n = b^m c^m$. Now, $b^p c^p = b^{m/n} c^{m/n} = (\sqrt[n]{b^m})(\sqrt[n]{c^m})$,
    so $(b^p c^p)^n = ((\sqrt[n]{b^m})(\sqrt[n]{c^m}))^n = (\sqrt[n]{b^m})^n (\sqrt[n]{c^m})^n = b^m c^m$. Hence $(bc)^p$
    and $b^p c^p$ have the same n^{th} power, so they are equal.

SECTION QUIZ

1. Eliminate the radicals from the denominator:

   (a) $(3\sqrt{5} + 2)/(\sqrt{5} - 1)$

   (b) $1/(\sqrt{x} + \sqrt{a})$, a is a constant.

   (c) $1/(\sqrt[3]{x} - \sqrt[3]{4})$

2. Simplify:

   (a) $36^{3/2} 6^{-3/2} / 6^{-1/2} \sqrt{9}$

   (b) $9^{-2} 3^{3/2} / (\sqrt{81})^2$

3. As part of the class assignment, a young couple was asked to simplify $[(x + 2)^2(y + x)^2]^3/(x^2 + xy + 2x + 2y)^5$. The young man suggests expanding the entire expression and then using long division. However, unknown to him, his girlfriend is secretly disguised as Super-Brain and shows him a much easier method. Explain the easier method and find the answer.

ANSWERS TO SECTION QUIZ

1. (a) $(17 + 5\sqrt{5})/4$

   (b) $(\sqrt{x} - \sqrt{a})/(x - a)$

   (c) $(\sqrt[3]{x^2} + \sqrt[3]{4x} + \sqrt[3]{16})/(x - 4)$

2. (a) 12

   (b) $3^{-13/2}$

3. Use the laws of exponents. The numerator is $[(x + 2)^2(y + x)^2]^3 = [((x + 2)(y + x))^2]^3 = ((x^2 + xy + 2x + 2y)^2)^3 = (x^2 + xy + 2x + 2y)^6$. Alternatively, you can factor the denominator: $(x^2 + xy + 2x + 2y)^5 = [x(x + y) + 2(x + y)]^5 = [(x + y)(x + 2)]^5$. The answer is $(x + y)(x + 2)$.

R.4   Straight Lines

PREREQUISITES

1.   There are no prerequisites for this section other than some high school

   analytic geometry; however, if the material presented in this section

   is new to you, it would be a good idea to enroll in a precalculus course.

   This section is intended to be a review.

GOALS

1.   Be able to find the distance between two points.

2.   Be able to write and manipulate equations of the line in its various

   forms.

3.   Be able to write the equation for perpendicular lines.

STUDY HINTS

1.   Distance formula. Don't be concerned about which point is (x_1, y_1)

   and which is (x_2, y_2) . The squaring process eliminates the need to

   make such a distinction. Remembering that the formula is derived

   from the Pythagorean theorem makes it easier to recall.

2.   Slope formula. As with the distance formula, don't be concerned

   about which point is (x_1, y_1) and which is (x_2, y_2) . The sign will

   correct itself when the division is performed.

3.   Point-slope form. Replacing (x_2, y_2) by (x, y) gives the slope as

   $m = (y - y_1)/(x - x_1)$. Rearrangement yields $y = y_1 + m(x - x_1)$.

4.   Slope - intercept form. Choosing $(x_1, y_1) = (0, b)$ in the point-slope

   form of the line yields $y = mx + b$.

5.   Point-point form. Substituting $m = (y_2 - y_1)/(x_2 - x_1)$ yields

   $y = y_1 + [(y_2 - y_1)/(x_2 - x_1)](x - x_1)$.

6. __Perpendicular lines.__ Many instructors try to write harder exam
problems by asking for equations of lines perpendicular to a given
line, so you will benefit by remembering that slopes of perpendicular
lines are negative reciprocals of each other, i.e., if a line has
slope m , then the perpendicular line has slope $-1/m$.

SOLUTIONS TO EVERY OTHER ODD EXERCISE

1.

For each point (x,y) , go x units
along the x-axis and the go y units
along the y-axis.

5. The distance from P_1 to P_2 is $\sqrt{(x_1 - x_2)^2 + (y_1 - y_2)^2}$. In
this case, the distance is $\sqrt{(1 - 1)^2 + (1 - (-1))^2} = \sqrt{4} = 2$.

9. The distance from P_1 to P_2 is $\sqrt{(x_1 - x_2)^2 + (y_1 - y_2)^2}$. In
this case, the distance is $\sqrt{(43721 - 3)^2 + (56841 - 56841)^2} =$
$\sqrt{(43718)^2} = 43718$.

13. The distance from P_1 to P_2 is $\sqrt{(x_1 - x_2)^2 + (y_1 - y_2)^2}$. In
this case, the distance is $\sqrt{(x - 3x)^2 + (y - (y + 10))^2} =$
$\sqrt{(-2x)^2 + (-10)^2} = \sqrt{4x^2 + 100} = 2\sqrt{x^2 + 25}$.

17. The slope of a line through (x_1, y_1) and (x_2, y_2) is
$m = (y_2 - y_1)/(x_2 - x_1)$. In this case, $m = (6 - 3)/(2 - 1) = 3$.

21.

The equation of the line with slope m
passing through (x_1, y_1) is

$y = y_1 + m(x - x_1)$. In this case,

$y = 3 + 2(x - 2) = 2x - 1$.

25. The equation of the line through (x_1, y_1) and (x_2, y_2) is

$y = y_1 + [(y_2 - y_1)/(x_2 - x_1)](x - x_1)$. In this case, $y = 7 +$
$[(4 - 7)/(-1 - 5)](x - 5) = 7 + (1/2)(x - 5) = x/2 + 9/2$ or
$2y = x + 9$.

29. We want to write the line in the form $y = mx + b$. Then m is the
slope and b is the y-intercept. $x + 2y + 4 = 0$ is the same as
$2y = -x - 4$ or $y = -x/2 - 2$. Thus, the slope is $-1/2$ and the
y-intercept is -2 .

33. Write the line in the form $y = mx + b$. Then m is the slope and
b is the y-intercept. $13 - 4x = 7(x + y) = 7x + 7y$ is equivalent
to $13 - 11x = 7y$ or $y = -(11/7)x + 13/7$. Thus, the slope is
$-11/7$ and the y-intercept is $13/7$.

37. (a) $4x + 5y - 9 = 0$ implies $5y = -4x + 9$, i.e., $y = -(4/5)x + 9/5$.
     This is in the form $y = mx + b$, so the slope is $-4/5$.

     (b) A perpendicular line has slope $5/4$. When this line passes
     through $(1,1)$, its equation is $y = 1 + (5/4)(x - 1) = (5x - 1)/4$
     or $4y = 5x - 1$.

41. Using the point-point form of the line, we get the equation
$y = 2 + [(4 - 2)/(2 - 4)](x - 4) = 2 + (-1)(x - 4) = -x + 6$.

SECTION QUIZ

1.   Sketch the line $3x + 2y = 1$.

2.   What is the slope of the line $5x - 8y = 4$?

3.   Find the equation of line passing through $(3,1/2)$ and perpendicular
     to the line going through $(98,3)$ and $(98,-10)$.

4.   One of your psychotic math friends has seen a little green space ship
     and a little red one land in her sink. According to her estimation,
     the green ship landed at $(-3,1)$ and the red one landed at $(2,4)$.
     She sees the aliens attacking each other with toothbrushes.   The
     sink is divided by the perpendicular passing through the midpoint
     between the spacecrafts. What is the equation of the line? What is
     the distance between the two ships?

ANSWERS TO SECTION QUIZ

1.

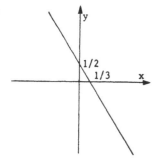

2.   5/8

3. $y = 1/2$

4.   Line: $3y + 5x + 5$; distance: $\sqrt{34}$.

R.5 Circles and Parabolas

PREREQUISITES

1.   There are no prerequisites for this section other than some high school
     analytic geometry; however, if the material presented in this section
     is new to you, it would be a good idea to enroll in a precalculus course.
     This section is intended to be a review.

GOALS

1.   Be able to recognize equations for circles and parabolas, and be able
     to describe their graphs.

2.   Be able to solve simultaneous equations to find intersection points.

STUDY HINTS

1.   Circles. The general equation is $(x - a)^2 + (y - b)^2 = r^2$, where
     (a,b) is the center of the circle and r is the radius. Any equation
     of the form $Ax^2 + Ay^2 + Bx + Cy + D = 0$ may be written in the general
     form of a circle by completing the square. If $r^2 > 0$, then the
     equation describes a circle. The coefficient of x^2 and y^2 must be
     the same.

2.   Parabolas. The general equation is $y - q = a(x - p)^2$, where (p,q)
     is the vertex. The parabola opens upward if $a > 0$, downward if
     $a < 0$. The graphs of all quadratic equations of the form $y = ax^2 +$
     $bx + c$ are parabolas.

3.   Simultaneous equations. The solutions represent points of intersection.
     One method of solution is to multiply equations by a constant factor and
     then subtract equations to eliminate a variable as in Example 6 . The
     other method uses substitution to eliminate a variable as in Example 7.

SOLUTIONS TO EVERY OTHER ODD EXERCISE

1.

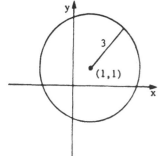

A circle with center at (a,b) and radius r has the equation $(x - a)^2 + (y - b)^2 = r^2$. With center (1,1) and radius 3, the equation is $(x - 1)^2 + (y - 1)^2 = 9$ or $x^2 - 2x + y^2 - 2y = 7$.

5.

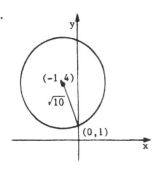

A circle with center at (a,b) and radius r has the equation $(x - a)^2 + (y - b)^2 = r^2$. With center at (-1,4), the equation is $(x +1)^2 + (y - 4)^2 = r^2$. Substituting in (0,1) yields $1^2 + (-3)^2 = r^2 = 10$, so the desired equation is $(x + 1)^2 + (y - 4)^2 = 10$ or $x^2 + 2x + y^2 - 8y + 7 = 0$.

9.

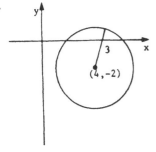

We complete the square to find the center and radius. $-x^2 - y^2 + 8x - 4y - 11 = 0$ is equivalent to $x^2 + y^2 - 8x + 4y + 11 = 0$, i.e., $(x^2 - 8x + 16) + (y^2 + 4y + 4) = -11 + 16 + 4 = 9 = (x - 4)^2 + (y + 2)^2$. Thus, the center is (4,-2) and the radius is 3.

13. The parabola with vertex at (p,q) has the general equation $y - q = a(x - p)^2$. In this case, $y - 5 = a(x - 5)^2$. Substituting (0,0) yields $-5 = a(-5)^2$, so $a = -1/5$ and the equation becomes $y = -(x - 5)^2/5 + 5 = -x^2/5 + 2x$.

17.

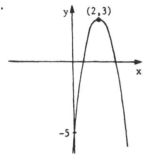

We complete the square to get the form

$y - q = a(x - p)^2$, where (p,q) is

the vertex. $y = -2x^2 + 8x - 5$ implies

$(y + 5) - 8 = -2(x^2 - 4x + 4) = -2(x - 2)^2 =$

$y - 3$. Thus, the vertex is $(2,3)$ and

since $a = -2 < 0$, the parabola opens

downward.

21.

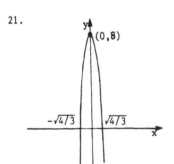

$y = -6x^2 + 8$ is the same as $y - 8 =$

$-6(x - 0)^2$. This has the form $y - q =$

$a(x - p)^2$, so the vertex is (p,q) or

$(0,8)$. Since $a = -6 < 0$, the parabola

opens downward.

25.

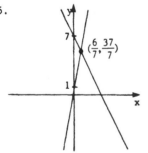

To find the point of intersection, solve

the equations simultaneously. $-2x + 7 =$

$y = 5x + 1$ implies $6 = 7x$, i.e., $x = 6/7$.

Substituting back into one of the equations,

we get $y = 5(6/7) + 1 = 37/7$, so the point

of intersection is $(6/7, 37/7)$.

To graph the equations, we see that

$y = -2x + 7$ is a line with slope -2 and y-intercept 7. $y = 5x + 1$ has

slope 5 and y-intercept 1.

29.

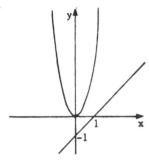

To find the point of intersection, solve the equations simultaneously. Thus, $y - x + 1 = 0$ becomes $3x^2 - x + 1 = 0$. The quadratic formula yields $x = (1 \pm \sqrt{1 - 12})/6$. Therefore, the parabola and the line do not intersect.

To graph the equations, we see that $y = 3x^2$ is an upward opening parabola with vertex at $(0,0)$. $y - x + 1 = 0$ is equivalent to $y = x - 1$, which is a line with slope 1 and y-intercept -1.

33.

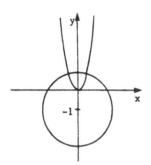

We solve the equations simultaneously. $y = 4x^2$ becomes $y/4 = x^2$, so $x^2 + 2y + y^2 - 3 = 0 = y/4 + 2y + y^2 - 3 = y^2 + 9y/4 - 3$. By the quadratic formula, $y = (-9/4 \pm \sqrt{81/6 + 12})/2 = (-9 \pm \sqrt{273})/8$. From $y = 4x^2$, we have $x = \pm\sqrt{y/4}$, and y must be non-negative. Thus, the inter-section points are $([(-9 + \sqrt{273})/32]^{1/2}, (-9 + \sqrt{273})/8)$ and $(-[(-9 + \sqrt{273})/32]^{1/2}, (-9 + \sqrt{273})/8)$.

$y = 4x^2$ is a parabola opening upward with vertex at the origin. $x^2 + 2y + y^2 - 3 = 0$ is equivalent to $x^2 + (y^2 + 2y + 1) = 4$ or $x^2 + (y + 1)^2 = 2^2$. This is a circle centered at $(0,-1)$ with radius 2.

37.

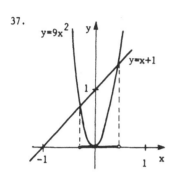

$9x^2 < x + 1$ implies $9x^2 - x - 1 < 0$.

The solution of $9x^2 - x - 1 = 0$ is

$(1 \pm \sqrt{1 + 36})/18$.   1/18   satisfies

$9x^2 < x + 1$,   and since   1/18   lies

between the solutions of $9x^2 - x - 1 = 0$,

the solution of the inequality is

$x \in ((1 - \sqrt{37})/18, (1 + \sqrt{37})/18)$.

Geometrically, the solution interval is where the line $x + 1$ lies
above the parabola $9x^2$.

SECTION QUIZ

1.   The curves $y - 1 = x^4$ and $y = 2x^2$ intersect at two points. Find them.

2.   They called her Melody the Marcher because she always got her employees
to march off to work. After working a day for the Marcher, you no longer
needed to be told, "Get to work!" Her glaring eyes told you, "You better
march back to work!" She wanted a circular fence of radius 7 to protect
a prized tree at (2,3) . One of the hired hands came up with a devious
plan. He decided to make a parabolic fence passing through (9,3),(-5,3)
and (2,-4) . After the fence was half finished, he asked her to stand at
the focus (see Fig. R.5.4). When the Marcher arrived, the other employees
threw garbage at the wall to reflect back at the focus.

(a) What is the equation of the parabola?

(b) What is the equation of the circle?

(c) Sketch both curves on the same set of axes.

ANSWERS TO SECTION QUIZ

1. $(-1,2),(1,2)$

2.   (a) $7y = x^2 - 4x - 24$

      (b) $x^2 - 4x + y^2 - 6y = 36$

      (c)

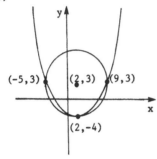

R.6   Functions and Graphs

PREREQUISITES

1.   There are no special prerequisites for this section; however, if the
     material presented in this section is new to you, it would be a good
     idea to enroll in a precalculus course.   This section is intended to
     be a review.

GOALS

1.   Be able to evaluate a function at a given point.

2.   Be able to sketch a graph by plotting.

3.   Be able to recognize the graph of a function.

STUDY HINTS

1.   Teminology.   Sometimes,   x   and   y   are referred to as the independent
     and dependent variables, respectively.   A good way to remember this is
     that normally   x   is chosen independently and then, the value of   y
     depends on   x .

2.   Calculator errors.   A calculator is accurate only up to a certain
     number of digits.   Round-off errors may accumulate, so be careful!
     Calculators do not always give the correct answer.

3.   Domain.   The domain of a function   f   is simply all of the   x   values
     for which   f(x)   is defined.   Note the emphasis on is; it is not can be .

4.   More on terminology.   Although many of us may get sloppy with the
     useage of   f ,   f(x) ,   and the graph of   f ,   you should know their
     distinctions.   f   is the function itself.   f(x)   refers to the func-
     tion value when the variable is x; the variable may just as well be any
     other letter.   Finally, the graph of   f   is a drawing which depicts   f .

5.   Graphing. For now, just be aware that plotting points at smaller
     intervals gives a more accurate drawing.

6.   Calculator plotting. A programmable calculator can be very useful
     for plotting Fig. R.6.8. Depending on your calculator, a variation
     of the following program may be used. $\boxed{\text{STO}}$ $\boxed{\text{RCL}}$ $\boxed{x^2}$ $\boxed{x^2}$ $\boxed{\times}$ $\boxed{0}$
     $\boxed{.}$ $\boxed{3}$ $\boxed{-}$ $\boxed{\text{RCL}}$ $\boxed{x^2}$ $\boxed{\times}$ $\boxed{0}$ $\boxed{.}$ $\boxed{2}$ $\boxed{-}$ $\boxed{0}$ $\boxed{.}$ $\boxed{1}$ $\boxed{=}$. x^2 is
     used twice to represent x^4 because most calculators will not raise
     negative numbers to a power other than two.

7.   Graphs of functions. Another way of recognizing graphs of functions
     is to retrace the curve starting from the left. If you do not have
     to go straight up or down, or go backwards to the left, then the
     graph depicts a function.

SOLUTIONS TO EVERY OTHER ODD EXERCISE

1. $f(-1) = 5(-1)^2 - 2(-1) = 7$; $f(1) = 5(1)^2 - 2(1) = 3$.

5. $f(-1) = -(-1)^3 + (-1)^2 - (-1) + 1 = 4$; $f(1) = -(1)^3 + (1)^2 - (1) + 1 = 0$.

9.   The domain of a function are the x-values for which there exists y-values.
     In this case, we need the expression under the radical sign to be non-
     negative. We want $1 - x^2 \geqslant 0$, so the domain is $-1 \leqslant x \leqslant 1$.
     $f(10)$ is not real.

13.

We complete the square as follows:

$$y = 5x^2 - 2x = 5(x^2 - 2x/5 + 1/25) - 1/5 =$$

$5(x - 1/5)^2 - 1/5$. This is a parabola

opening upwards with vertex at $(1/5, -1/5)$.

17.

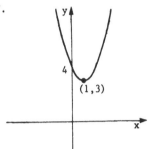

$y = (x - 1)^2 + 3$ is a parabola opening

upwards with vertex at $(1,3)$.

21.

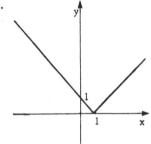

$y = |x - 1|$ is $y = x - 1$ if $x - 1 \geqslant 0$,

i.e., if $x \geqslant 1$.   Also, $y = |x - 1|$ is

$y = -(x - 1) = -x + 1$ if $x - 1 < 0$,   i.e.,

if $x < 1$.   Thus, the graph consists of two

line segments.

25.

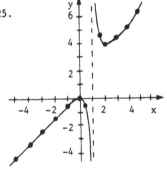

The entire graph is shown at the left.   The

10 points should lie on the graph.   Some

arbitrary points include: $(-10,-9.1)$,

$(-8,7.1)$, $(-6,-5.1)$, $(-4,-3.2)$, $(-2,1.3)$,

$(-1, -0.5)$, $(0,0)$, $(0.5,-0.5)$, $(1.5,4.5)$,

$(3,4.5)$, $(5,6.25)$,   and $(6,7.2)$.

29.   A graph of a function has only one y-value for any given x-value, i.e.,

each vertical line intersects the graph at only one point.   Thus, only

(a)   and   (c)   represent functions.   (d)   is not a function due to the

vertical line segment in the graph.

33. $y = \sqrt{x^2 - 1}$ has only one value for any given x ,   so it is a function.

We want $x^2 - 1 \geqslant 0$,   i.e., $x \leqslant -1$ and $x \geqslant 1$ make up the domain.

SECTION QUIZ

1. Suppose $g(x) = x^3 - x + 1$. What is $g(-1)$? $g(0)$? $g(y)$?

2. What is the domain of f if $f(x) = \begin{cases} 1/(1 - x) & x < 0 \\ x^2 & x \geq 0 \end{cases}$?

3. What is the domain of f if $f(x) = \begin{cases} 1/(1 - x) & x > 0 \\ x^2 & x \leq 0 \end{cases}$?

4. In Questions 2 and 3, which are functions?

5. In the midst of a nightmare, your roommate believes a giant black mouse is chasing him. Sometimes, he tries to fool the mouse by running backwards. Thus, when he awakens in a cold sweat, his position from the point of origin is given by $f(x) = 36x - x^2$.

    (a) If $f(x)$ must be nonnegative, what is the domain of f ?

    (b) Sketch $f(x)$, given the domain in part (a).

ANSWERS TO SECTION QUIZ

1. $1, 1$, and $y^3 - y + 1$.

2. $(-\infty, \infty)$.

3. $x \neq 1$.

4. Both

5. (a) $x \in [0, 36]$

    (b)
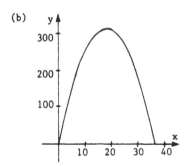

R.R   Review Exercises for Chapter R

SOLUTIONS TO EVERY OTHER ODD EXERCISE

1.   Subtract 2 from both sides to get $3x = -2$. Then divide by 3 to
     get $x = -2/3$.

5.   Expansion gives $(x + 1)^2 - (x - 1)^2 = 2 = (x^2 + 2x + 1) - (x^2 - 2x + 1) =$
     $4x$. Divide both sides by 4 to get $x = 1/2$.

9. $8x + 2 > 0$ is equivalent is $8x > -2$. Divide both sides by 8 to get
     $x > -1/4$.

13. Expand to get $x^2 - (x - 1)^2 = x^2 - (x^2 - 2x + 1) = 2x - 1 > 2$. Add 1
     to both sides to get $2x > 3$ and divide by 2 to get $x > 3/2$.

17. $x^2 < 1$ is equivalent to $x^2 - 1 < 0$ or $(x + 1)(x - 1) < 0$. One possible
     solution is $x < -1$ and $x > 1$, but this is not possible. Another
     possibility is $x > -1$ and $x < 1$. Thus, the solution is $-1 < x < 1$,
     i.e., $x \in (-1,1)$.

21. $|x - 1|^2 \geqslant 2$ means $|x - 1| \geqslant \sqrt{2}$, i.e., $x - 1 \geqslant \sqrt{2}$ or $x - 1 \leqslant -\sqrt{2}$.
     For $x - 1 \geqslant \sqrt{2}$, we get $x \geqslant 1 + \sqrt{2}$. For $x - 1 \leqslant -\sqrt{2}$, we get
     $x \leqslant -\sqrt{2} + 1$. Therefore, the solution is $x \geqslant 1 + \sqrt{2}$ or $x \leqslant 1 - \sqrt{2}$,
     i.e., $x \in (-\infty , 1 - \sqrt{2}]$ or $x \in [1 + \sqrt{2} , \infty)$.

25. $x^3 \in (-8,27)$ implies $x \in (-2,3)$, so $x < 10$ and $x \in (-2,3)$ means
     $-2 < x < 3$; i.e., $x \in (-2,3)$.

29. $2(7 - x) \geqslant 1$ is equivalent to $14 - 2x \geqslant 1$ or $-2x \geqslant -13$. Dividing
     by -2 reverses the inequality and yields $x \leqslant 13/2$. Also, $3x - 22 > 0$
     is equivalent to $3x > 22$ or $x > 22/3$. Thus, the solution is
     $x \in (-\infty,13/2]$ or $x \in (22/3,\infty)$.

33. $|-8| = 8$, so $|-8| + 5 = 13$.

37. $\sqrt{2} \cdot 2^{-1/2} = \sqrt{2}/2^{1/2} = \sqrt{2}/\sqrt{2} = 1$.

41. $x^{1/4} \sqrt{x} \, y/x^{1/2} \, y^{3/4} = x^{1/4+1/2-1/2} \, y^{1-3/4} = x^{1/4} \, y^{1/4} = \sqrt[4]{xy}$.

45. The distance between P_1 and P_2 is $\sqrt{(x_2 - x_1)^2 + (y_2 - y_1)^2}$.
     In this case, it is $\sqrt{(2 - (-1))^2 + (0 - 1)^2} = \sqrt{9 + 1} = \sqrt{10}$.

49. The point-point form of a line is $y = y_1 + [(y_2 - y_1)/(x_2 - x_1)(x - x_1)]$.
     In this case, the line is $y = -1 + [(3 - (-1))/(7 - 1/2)](x - 1/2) =$
     $-1 + [4/(13/2)](x - 1/2) = -1 + (8/13)(x - 1/2) = 8x/13 - 17/13$, i.e.,
     $13y - 8x + 17 = 0$.

53. The point-slope form of a line is $y = y_1 + m(x - x_1)$. In this case, the
     line is $y = 13 + (-3)(x - 3/4) = -3x + 61/4$ or $4y + 3x = 61$.

57. The slope of $5y + 8x = 3$ is $-8/5$. Thus, the slope of a perpendicular
     line is $5/8$, and the line passing through $(1,1)$ is $y = 1 + (5/8)(x - 1)$,
     using point-slope form of the line. The line is $y = 5x/8 + 3/8$ or
     $8y - 5x = 3$.

61. The equation of the circle with center (a,b) and radius r is
     $(x - a)^2 + (y - b)^2 = r^2$. In this case, the circle is $(x - 12)^2 +$
     $(y - 5)^2 = 8^2$ or $(x^2 - 24x + 144) + (y^2 - 10y + 25) = 64$ or
     $x^2 - 24x + y^2 - 10y + 105 = 0$.

65.

Complete the square to get

$y = 3(x^2 + 4x/3 + 4/9) + 2 - 4/3 =$

$3(x + 2/3)^2 + 2/3$. This is a parabola

opening upward with vertex at $(-2/3, 2/3)$.

69. Solve the equations simultaneously. Substitute $y = x$ into $x^2 + y^2 = 4$
     to get $2x^2 = 4$, i.e., $x^2 = 2$ or $x = \pm\sqrt{2}$. Thus, the points of inter-
     section are $(-\sqrt{2}, -\sqrt{2})$ and $(\sqrt{2}, \sqrt{2})$.

73.

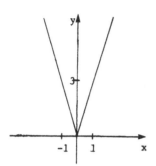

This is the graph of $y = 3x$ if $x \geqslant 0$;
it is $y = -3x$ if $x < 0$.

77. (a)

$f(-2) = -0.6$; $f(0) = 0$;
$f(2) = 0.6$.

(b)

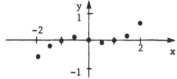

In addition to the points in
part (a), we have $f(-1.5) =$
-0.1875 ; $f(-1) = 0$; $f(-0.5) =$
0.0375 ; $f(0.5) = -0.0375$;

$f(1) = 0$; $f(1.5) = 0.1875$.

81. If a vertical line passes through two points of the graph, it is not a
function. Thus, (a) and (c) are functions.

TEST FOR CHAPTER R

1. True or false:

(a) If $a > b > 0$, then $1/a > 1/b$.

(b) The interval $(2,4)$ contains 3 integers.

(c) If $a > b > 0$, then $ac > bc$ for constant c .

(d) The domain of $\sqrt{x}$ is $x \geqslant 0$.

(e) The line $x = 2$ has slope zero.

2. Express the solution set of $x^2 + 3x + 2 \geqslant 0$ in terms of absolute values.

3. Write equations for the following lines:

    (a) The line going through $(1,3)$ and $(2,4)$.

    (b) The line with slope 5 and passing through $(-3,2)$.

    (c) The line with slope -1 and y-intercept $1/2$.

4. Sketch the graph of $y = ||2x| - 2|$.

5. Do the following equations describe a circle or a parabola?

    (a) $x^2 + y^2 + 5x - 4y - 6 = 0$.

    (b) $y = x^2 - 2x + 3$.

    (c) $x + y^2 = 0$.

6. (a) Complete the square for the equation $y^2 + 4y + 3 = 0$.

    (b) Solve the equation in part (a) .

7. Sketch the graph of $y = |x^2 - 1|$.

8. (a) Find the points of intersection of the graphs of $y = x^2$ and $y = 5x - 6$.

    (b) Find the distance between the intersection points.

9. Factor the following expressions.

    (a) $x^2 + 2x - 15$

    (b) $x^3 - xy^2$

10. Dumb Donald had heard how wonderful chocolate mousse tasted, so he decided to go on a hunting trip at Moose Valley. When he finally spotted a moose, he chased the moose in circles around a tree located at $(-4,-2)$. Unfortunately for Dumb Donald, the moose ran much faster, caught up with Dumb Donald, and trampled him. Their path was radius 7 from the tree. What is the equation of the circle?

ANSWERS TO CHAPTER TEST

1. (a) False; $1/a < 1/b$.

    (b) False; 2 and 4 are excluded from the interval.

    (c) False; c must be positive for $ac > bc$.

    (d) True.

    (e) False; vertical lines do not have slopes.

2. $|x - 3/2| \geqslant 1/2$.

3. (a) $y = x + 2$.

    (b) $y = 5x + 17$.

    (c) $y = -x + 1/2$.

4.

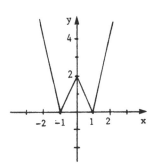

5. (a) Circle

    (b) Parabola

    (c) Parabola

6. (a) $(y + 2)^2 - 1 = 0$.

    (b) $y = -3$ or $y = -1$.

7.

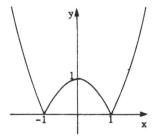

8.   (a)   (2,4)   and   (3,9)

     (b) $\sqrt{26}$

9.   (a) $(x + 5)(x - 3)$

     (b) $x(x + y)(x - y)$

10. $x^2 + y^2 + 8x + 4y = 29$.

.

CHAPTER 1

DERIVATIVES AND LIMITS

1.1 Introduction to the Derivative

PREREQUISITES

1.   Recall how to use functional notation (Section R.6).

2.   Recall how to graph straight lines and how to compute the slopes of
     the lines (Section R.4).

PREREQUISITE QUIZ

1.   Let $f(x) = x^2 + 3x + 4$. Evaluate the following expressions:

     (a) $f(3)$

     (b) $f(y)$

     (c) $f(x + y)$

2.   Find the slopes of the lines given by the following equations:

     (a) $2x + 2y = 5$

     (b) $y = -x + 3$

3.   Sketch the graphs of the equations given in Question 2.

GOALS

1.   Be able to calculate the velocity of an object from its position formula.

2.   Be able to compute the slope of the tangent line for a given curve.

3.   Be able to reproduce and understand the formula

$$f'(x_0) = \lim_{\Delta x \to 0} \frac{\Delta y}{\Delta x} = \lim_{\Delta x \to 0} \frac{f(x_0 + \Delta x) - f(x_0)}{\Delta x} \; .$$

STUDY HINTS

1.   <u>Definitions</u>. Average velocity is the ratio of distance travelled to
     time elapsed over a period of time. Instantaneous velocity is the
     same ratio as the time period approaches zero. In more general sit-
     uations, average velocity corresponds to average rate of change and
     instantaneous velocity corresponds to the instantaneous rate of change.

2.   <u>Definition of derivative</u>. Know that $\Delta y / \Delta x$ for Δx approaching zero
     is the derivative. Also, notice that $\Delta y = f(x + \Delta x) - f(x)$.

3.   <u>Cancelling Δx</u> . If, in working out $\Delta y / \Delta x$, Δx can not be cancelled
     from the denominator, you should suspect a computational error in most
     cases. Be sure you know how to use the definition to find the derivative.

4.   <u>Physical interpretation</u>. Instantaneous velocity and slope of the tan-
     gent line are some of the many names that the derivative specializes to
     in specific situations.

5.   <u>Notation</u>. Familiarize yourself with $f'(x)$.

6.   <u>Quadratic function rule</u>. Memorize the formula for now. A more general
     formula will be memorized in Section 1.4 which will subsume this one.
     Note what happens if $a = 0$ or if a and b are both 0 ; the linear
     and constant function rules result.

SOLUTIONS TO EVERY OTHER ODD EXERCISE

1. By definition, $\Delta y = f(x_0 + \Delta x) - f(x_0)$ and the average velocity is $\Delta y / \Delta x$.

(a) $\Delta y = f(2 + 0.5) - f(2) = [(2.5)^2 + 3(2.5)] - [(2)^2 + 3(2)] =$
13.75 - 10 = 3.75 ; $\Delta y / \Delta x = 3.75/0.5 = 7.5$.

(b) $\Delta y = f(2 + 0.01) - f(2) = [(2.01)^2 + 3(2.01)] - [(2)^2 + 3(2)] =$
10.0701 - 10 = 0.0701 ; $\Delta y / \Delta x = 0.0701/0.01 = 7.01$.

(c) $\Delta y = f(4 + 0.1) - f(4) = [(4.1)^2 + 3(4.1)] - [(4)^2 + 3(4)] =$
29.11 - 28 = 1.11 ; $\Delta y / \Delta x = 1.11/0.1 = 11.1$.

(d) $\Delta y = f(4 + 0.01) - f(4) = [(4.01)^2 + 3(4.01)] - [(4)^2 + 3(4)] =$
28.1101 - 28 = 0.1101 ; $\Delta y / \Delta x = 0.1101/0.01 = 11.01$.

5. The instantaneous velocity is given by $f'(x_0)$. By the quadratic function rule, $f'(x) = 2x + 3$; therefore, $f'(2) = 7$ m/sec.

9. (a) The instantaneous velocity is given by $f'(x_0)$. By the quadratic function rule, $f'(x_0) = 2x_0 + 3$.

(b) The instantaneous velocity is 10 m/sec when $2x_0 + 3 = 10$, i.e., $x_0 = 7/2$ seconds.

13.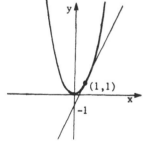

The slope of the tangent line at x_0 is $f'(x_0)$. We begin by finding $\Delta y / \Delta x = [(1 + \Delta x)^2 - (1)^2]/\Delta x = [2\Delta x + (\Delta x)^2]/\Delta x = 2 + \Delta x$. As Δx gets small, we get $f'(1) = 2$.

17. By definition, $f'(x_0) = \Delta y / \Delta x$ for a small Δx . Thus, $f'(x_0) = [f(x_0 + \Delta x) - f(x_0)]/\Delta x = [(a\Delta x + 2) - 2]/\Delta x = a\Delta x / \Delta x = a$.

21. The quadratic function rule states that the derivative of $ax^2 + bx + c$ is $2ax + b$. Here, $a = 1$, $b = 1$, and $c = -1$, so the derivative is $2x + 1$. At $x_0 = 1$, $f'(1) = 3$.

25. (a) The vertex of the parabola is the point where the derivative is

     zero. By the quadratic function rule, $f'(x) = 2x - 16$, which

     is zero at $x_0 = 8$. Substitution gives $y = (8)^2 - 16(8) + 2 =$

     -62 , so the desired point is $(8,-62)$.

   (b) Algebraically, the vertex occurs at the highest or lowest point,

     which may be found by completing the square. $y = (x^2 - 16x + 64) +$

     $(2 - 64) = (x - 8)^2 - 62$. $(x - 8)^2$ is zero for $x = 8$ and

     positive otherwise, so the parabola has its lowest point at $x = 8$.

29. By the quadratic function rule, the derivative of $ax^2 + bx + c$ is

     $2ax + b$. Here, $a = 1$, $b = 3$, and $c = -1$, so $f'(x) = 2x + 3$.

33. By the quadratic function rule, the derivative of $at^2 + bt + c$ is

     $2at + b$. Here, $a = -4$, $b = 3$, and $c = 6$, so $g'(t) =$

     $-8t + 3$.

37. As in Example 8, the best time to make the jump is when the velocities

     are equal. We want to equate the derivatives: $4t + 3 = 6t + 1$.

     Therefore, the jump should be made at $t = 1$.

41. Here, $f'(x) = 2x$, so $f'(3) = 6$.

45. Expanding, $k(y) = y^2 - 3y - 28$, so $k'(y) = 2y - 3$ and $k'(-1) = -5$.

49. $\Delta y/\Delta x = \{ [4(x + \Delta x)^2 + 3(x + \Delta x) + 2] - [4x^2 + 3x + 2]\}/\Delta x =$

     $(8x\Delta x + 4(\Delta x)^2 + 3\Delta x)/\Delta x = 8x + 3 + 4\Delta x$. As Δx gets small, the

     derivative becomes $8x + 3$.

53. $\Delta y/\Delta x = \{ [-2(x + \Delta x)^2 + 5(x + \Delta x)] - [-2x^2 + 5x]\}/\Delta x = \{-4x\Delta x - 2(\Delta x)^2 +$

     $5\Delta x\}/\Delta x = -4x-2\Delta x + 5$. As Δx gets small, the derivative becomes $-4x$ ·

57. From basic geometry, $A(x) = x^2$ and the perimeter $P(x) = 4x$. $A'(x) =$

     $2x$, which is $(1/2)P(x)$.

61.

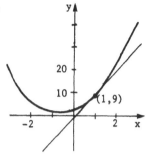

By the quadratic function rule, $f'(x) =$ $6x + 4$, so $f'(1) = 10$. Since $f(1) =$ 9, the point-slope form of the line yields $y = 9 + 10(x - 1)$ or $y = 10x - 1$. By Example 7, the vertex of a parabola occurs where $f'(x_0) = 0$. In this case $6x + 4 = 0$ at $x = -2/3$.

65.

If $f(x) = x^2$, $f'(x) = 2x$, so the slope of the tangent line is $2R$; therefore, the perpendicular line is $y = R^2 + \left(-\dfrac{1}{2R}\right)(x - R)$. This line crosses $x = 0$ at $y = R^2 + 1/2$. The horizontal crosses at $y = R^2$, so the distance is $1/2$.

69. (a) Let $f(t) = 24.5t - 4.9t^2$. The velocity at time t is $f'(t)$. By the quadratic function rule, $f'(t) = 2(-4.9)t + 24.5$, i.e., $f'(t) = -9.8t + 24.5$. Substituting different values of t, we have the velocity at $t = 0,1,2,3,4,5$ equal to $24.5, 14.7, 4.9, -4.9, -14.7, -24.5$ meters/sec, respectively.

(b) $f(t) = 24.5t - 4.9t^2 = -4.9(t - 2.5)^2 + 30.625$ is a parabola facing downward. The highest point is at the vertex when $t = 2.5$ sec.

(c) The velocity is equal to zero when $-9.8t + 24.5 = 0$, i.e., $t = 2.5$ sec. This result is consistent with part (b).

(d) The ball strikes the ground when $f(t) = 0$. Solving for $24.5t - 4.9t^2 = 0$ gives $t = 5$ sec.

SECTION QUIZ

1. If a particle's position is given by $y = x^2 + 5x - 7$ after x seconds, compute the particle's velocity at $y_0 = -1$. (Hint: there are two answers.)

2. Compute $f'(x)$:

    (a) $f(x) = x^2 + x + 1$

    (b) $f(x) = 3\pi^2$

3. Compute $\Delta y/\Delta x$ for $y = x^2/4 + 3x/2 + 2.7$.

4. Find the slope of the tangent line of the parabola in Question 3 at $x_0 = 4$.

5. Just a little while ago, the notorious Claude Crum ran off with sweet Cindy Lou, the southern belle daughter of Deputy Dan. Claude's position is $5x + 3$ miles after x hours. Deputy Dan's horse, who takes superduper vitamins, can run faster, so its position is $3x^2 + 2x$ at time x hours.

    (a) Find the speed of both individuals at any time x .

    (b) When will Deputy Dan catch up with Claude?

    (c) What is the speed of both individuals when Dan catches Claude and stops sweet Cindy Lou from marrying Claude?*

---

*Dear Reader: I realize that many of you hate math but are forced to complete this course for graduation. Thus, I have attempted to maintain interest with "entertaining" word problems. They are not meant to be insulting to your intelligence. Obviously, most of the situations will never happen; however, calculus has several practical uses and such examples are found throughout Marsden and Weinstein's text. I would appreciate your comments on whether my "unusual" word problems should be kept for the next edition.

ANSWERS TO PREREQUISITE QUIZ

1.   (a)   22

     (b) $y^2 + 3y + 4$

     (c) $(x + y)^2 + 3(x + y) + 4$

2.   (a)   -1

     (b)   -1

3.   (a)   (b)

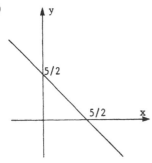   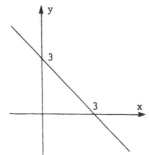

ANSWERS TO SECTION QUIZ

1. $y'(1) = 7$ and $y'(-6) = -7$

2.   (a) $2x + 1$

     (b)   0

3. $x/2 + 3/2 + \Delta x/4$

4.   7/2

5.   (a)   Claude: 5miles/hour; Dan: $(6x + 2)$miles/hour

     (b) $[(1 + \sqrt{5})/2]$ hours

     (c)   Claude: 5miles/hour; Dan: $(5 + 3\sqrt{5})$miles/hour

1.2 Limits

PREREQUISITES

1. Recall how to factor simple polynomials. (Section R.1)

2. Recall that graphs may be used to depict functions. (Section R.6)

PREREQUISITE QUIZ

1. Factor the following:

    (a) $x^4 - y^4$

    (b) $x^3 - 8$

    (c) $4x^2 + 16x + 16$

    (d) $3x^2 + 8x + 4$

2. Sketch the graphs of the following functions:

    (a) $y = -x + 2$

    (b) $y = x^3 - 1$

GOALS

1. Be able to find limits by computation and by reading graphs.

STUDY HINTS

1. Cautions about limits. $\lim\limits_{x \to a} f(x)$ depends on what happens around $x = a$,
    not what happens at $x = a$. Also, the limiting value must be approached
    from both sides of $x = a$, not just one side. Example 2 demonstrates
    these points well.

2. **Basic properties of limits**. Know how to use these statements for com-
    puting limits. Most of them are intuitively obvious, so there is no need
    to memorize. You should take a few moments to convince yourself that
    these statements are correct.

3. **Replacement rule**. This rule merely states that limits of equivalent
    functions are equal. The method of Example 3(b) is commonly used on
    test questions. The idea is to cancel common factors to leave a nonzero
    denominator. Usually, one of the factors is $(x - a)$ if we want to find
    the limit as x approaches a (See Example 5(a)).

4. **Continuity defined**. This is a very important concept in mathematics.
    Notice that $\lim_{x \to x_0} f(x)$ must exist and that it must equal $f(x_0)$ if
    f is continuous at x_0.

5. **Rational functions**. These functions, polynomials or quotients of poly-
    nomials, are continuous at x_0 if the denominator does not equal zero.
    Their limits may be computed at x_0 by "substituting" $x = x_0$; however,
    be sure the function is rational before applying this rule.

6. **Limits of infinity**. Example 8(a) shows the important fact that
    $\lim_{x \to \pm\infty} (1/x) = 0$. To compute limits at infinity, divide numerator and
    denominator by an appropriate power of x so that all except for one
    term in the denominator approaches zero. See Example 8, parts (b) and (c).

7. **Terminology**. "Limit does not exist" can either mean the limit is approach-
    ing infinity or that no particular value is being approached as x ap-
    proaches x_0.

8.   <u>One-sided limit</u>. Notice that in Example 7, no limit exists at zero. However, as we approach zero from either the left or the right, the function does approach a certain value. This is the concept of one-sided limits. More details will come later in this course.

SOLUTIONS TO EVERY OTHER ODD EXERCISE

1.   Let $f(x) = (x^3 - 3x^2 + 5x - 3)/(x - 1)$. We find $f(1.1) = 2.01$, $f(1.001) = 2.000001$, $f(0.9) = 2.01$, $f(0.999) = 2.000001$. These results suggest that $\lim\limits_{x \to 1} f(x) = 2$. To verify this, we divide the numerator by $x - 1$, obtaining $f(x) = [(x - 1)(x^2 - 2x + 3)]/(x - 1)$ which, for $x \neq 1$, is equal to $x^2 - 2x + 3$. Applying the replacement rule and the continuity of rational functions, we have $\lim\limits_{x \to 1} f(x) = \lim\limits_{x \to 1}(x^2 - 2x + 3) = 1^2 - 2 \cdot 1 + 3 = 2$.

5.   Using the method of Example 3, we have $\lim\limits_{x \to 3} (17 + x) = \lim\limits_{x \to 3} 17 + \lim\limits_{x \to 3} x$ (sum rule) $= 17 + 3$ (constant function and identity function rules) $= 20$

9.   Using the method of Example 4, we have $\lim\limits_{x \to 3} [(x^2 - 9)/(x^2 + 3)] =$ $\lim\limits_{x \to 3} (x^2 - 9) / \lim\limits_{x \to 3} (x^2 + 3)$ (quotient rule) $= \left[\lim\limits_{x \to 3} x^2 + \lim\limits_{x \to 3} (-9) \right] / \left[\lim\limits_{x \to 3} x^2 + \lim\limits_{x \to 3} 3 \right]$ (sum rule) $= (9 - 9)/(9 + 3)$ (power and constant function rule) $= 0$.

13. First, we note that $(u - \sqrt{3})/(u^2 - 3) = (u - \sqrt{3})/(u - \sqrt{3})(u + \sqrt{3}) = 1/(u + \sqrt{3})$. By the replacement rule, $\lim\limits_{u \to \sqrt{3}} [(u - \sqrt{3})/(u^2 - 3)] = \lim\limits_{u \to \sqrt{3}} [1/(u + \sqrt{3})]$. By the continuity of rational functions, the limit is $1/2\sqrt{3} = \sqrt{3}/6$.

17. We note that $f(x) = (x^2 - 4x + 3)/(x^2 - 2x - 3) = (x - 3)(x - 1)/(x - 3)(x + 1) = (x - 1)/(x + 1)$; therefore, by the replacement rule, $\lim\limits_{x \to 3} f(x) = \lim\limits_{x \to 3} [(x - 1)/(x + 1)]$. Then, by the continuity of rational functions, the limit becomes $2/4 = 1/2$.

21. Note that $f(\Delta x) = [3(\Delta x)^2 + 2(\Delta x)]/\Delta x = 3(\Delta x) + 2$; therefore, by the replacement rule, $\lim_{\Delta x \to 0} f(\Delta x) = \lim_{\Delta x \to 0} (3(\Delta x) + 2) = 2$.

25. Since $\sqrt{x}$ is continuous, we have $\lim_{x \to 3} [(1 - \sqrt{x})(2 + \sqrt{x})] = \lim_{x \to 3} (2 - \sqrt{x} - x) = 2 - \sqrt{3} - 3 = -\sqrt{3} - 1$.

29. $|x - 1|$ is always positive. If $x > 1$, then $|x - 1|/(x - 1) = 1$.

   If $x < 1$, then $|x - 1|/(x - 1) = -1$. Since the function approaches

   -1 from the left and it approaches 1 from the right, no limit exists.

33. Divide by x/x , so $\lim_{x \to -\infty} [(2x - 1)/(3x + 1)] = \lim_{x \to -\infty} [(2 - 1/x)/(3 + 1/x)] = (2 - 0)/(3 + 0) = 2/3$. We used the fact $\lim_{x \to -\infty} (1/x) = 0$.

37. $\lim_{x \to 0} f(x) = 1 = f(0)$. $\lim_{x \to 1} f(x) = 2 \neq f(1)$, $f(1)$ is not defined.

   $\lim_{x \to 2} f(x)$ is not defined while $f(2) = 2$ is defined. $\lim_{x \to 3} f(x) = 1 \neq f(3)$,

   $f(3)$ is not defined. $\lim_{x \to 4} f(x)$ does not exist since no finite number is

   approached as x approaches 4 , while $f(4) = 2$.

41. By the replacement rule, $\lim_{x \to 0} (-x/x^3) = -\lim_{x \to 0} (1/x^2) = -\infty$. This is

   because $1/x^2$ is always positive.

45. By the replacement rule, $\lim_{u \to 0} [(u^3 + 2u^2 + u)/u] = \lim_{u \to 0} (u^2 + 2u + 1) = 1$.

49. Note that $f(x) = (x^2 - 5x + 6)/(x^2 - 6x + 8) = (x - 3)(x - 2)/$

   $(x - 4)(x - 2) = (x - 3)/(x - 4)$. By the replacement rule, $\lim_{x \to 2} f(x) = \lim_{x \to 2} [(x - 3)/(x - 4)] = (-1)/(-2) = 1/2$.

53. Expansion yields $\lim_{\Delta x \to 0} [(125 + 75\Delta x + 15(\Delta x)^2 + (\Delta x)^3 - 125)/\Delta x]$, and

   the replacement rule gives $\lim_{\Delta x \to 0} (75 + 15\Delta x + (\Delta x)^2) = 75$.

57. Factoring out $s - 3$ yields $\lim_{s \to 3} [(3s + 7)/(s - 3)]$. Since $s < 3$

   causes the function to approach $-\infty$ and $s > 3$ causes the function to

   approach $+\infty$, no limit exists.

61. (a) If the block of ice melts completely at time T , there is no ice,
    so neither the base nor the height exists, so $f(T)$ and $g(T)$ must
    both equal zero.

    (b) The heat distribution need not be uniform throughout the room even
    though temperature is held constant. It could be the case that just
    before melting is completed, a very thin sheet of ice of area A is
    left. Then, $\lim\limits_{t\to T} g(t) = 0 = g(T)$, but $\lim\limits_{t\to T} f(t) = A \neq f(T)$.
    Another possibility is melting along the side faces only, with no
    melting in a vertical direction. The result is a tiny pillar of ice
    of height h just before melting is complete. Then, $\lim\limits_{t\to T} f(t) = 0 =$
    $f(T)$, but $\lim\limits_{t\to T} g(t) = h \neq g(T)$.

    (c) Volume is given by the formula $f(t)\cdot g(t)$, so $\lim\limits_{t\to T} [f(t)\cdot g(t)] = 0$

    (d) The product rule for limits states: $\lim\limits_{t\to T} [f(t)\cdot g(t)] = \lim\limits_{t\to T} f(t)\cdot$
    $\lim\limits_{t\to T} g(t)$. The first example in part (b) gives the following resul!
    $\lim\limits_{t\to T} f(t)\cdot\lim\limits_{t\to T} g(t) = A\cdot 0 = 0 = \lim\limits_{t\to T} [f(t)\cdot g(t)]$. Similarly, the secon‹
    example in part (b) gives: $\lim\limits_{t\to T} f(t)\cdot\lim\limits_{t\to T} g(t) = 0\cdot h = 0 = \lim\limits_{t\to T} [f(t)\cdot g$

65. (a) Having shown that $\lim\limits_{x\to x_0} [f_1(x) + f_2(x) + f_3(x)] = \lim\limits_{x\to x_0} f_1(x) +$
    $\lim\limits_{x\to x_0} f_2(x) + \lim\limits_{x\to x_0} f_3(x)$, we let $g(x) = f_2(x) + f_3(x) + f_4(x)$.
    Now, $\lim\limits_{x\to x_0} [f_1(x) + g(x)] = \lim\limits_{x\to x_0} f_1(x) + \lim\limits_{x\to x_0} g(x)$ by the basic
    sum rule. By using the extended sum rule for $n = 3$ on $\lim\limits_{x\to x_0} g(x)$,
    we obtain our desired result.

    (b) We let $g(x) = f_2(x) + f_3(x) + \ldots + f_{17}(x)$. Applying the extended
    sum rule for $n = 16$, we have $\lim\limits_{x\to x_0} g(x) = \lim\limits_{x\to x_0} f_2(x) +$
    $\lim\limits_{x\to x_0} f_3(x) + \ldots + \lim\limits_{x\to x_0} f_{17}(x)$. By the basic sum rule,
    $\lim\limits_{x\to x_0} [f_1(x) + f_2(x) + \ldots + f_{17}(x)] = \lim\limits_{x\to x_0} [f_1(x) + g(x)] =$
    $\lim\limits_{x\to x_0} f_1(x) + \lim\limits_{x\to x_0} g(x)$, and substitution gives the desired
    result.

65 (continued).

(c) Now, let $g(x) = f_2(x) + f_3(x) + \ldots + f_{n+1}(x)$. Apply the extended

sum rule for $n \geqslant 2$, we have $\lim\limits_{x \to x_0} g(x) = \lim\limits_{x \to x_0} f_2(x) =$

$\lim\limits_{x \to x_0} f_3(x) + \ldots + \lim\limits_{x \to x_0} f_{n+1}(x)$. By the basic sum rule,

$\lim\limits_{x \to x_0} [f_1(x) + f_2(x) + \ldots + f_{n+1}(x)] = \lim\limits_{x \to x_0} [f_1(x) + g(x)] =$

$\lim\limits_{x \to x_0} f_1(x) + \lim\limits_{x \to x_0} g(x)$. Substitution shows that the extended

sum rule holds for $n + 1$.

(d) In part (c), the statement holds for $n + 1$ assuming it holds for

n . Earlier, we showed that the extended sum rule holds for $m = 3$

therefore, the extended sum rule is proven by induction.

SECTION QUIZ

1. Compute the limit:

(a) $\lim\limits_{x \to 3} |(x - 3)/(x - 3)|$

(b) $\lim\limits_{x \to 3} [(x - 3)/(x + 3)]$

(c) $\lim\limits_{x \to 3} ((x - 3)/|x - 3|)$

2. Find:

(a) $\lim\limits_{x \to 2} [(x - 2)/(x^3 - 8)]$

(b) $\lim\limits_{x \to \infty} [(x - 2)/(x^3 - 8)]$

(c) $\lim\limits_{x \to 1} [(x^2 - 1)/(2x^2 + x - 3)]$

(d) $\lim\limits_{x \to \infty} [(x^2 - 1)/(2x^2 + x - 3)]$

3. Let $f(x) = 2x^3 + 6x + 7$. Compute $f'(3)$.

4.   Sir Chuck, the clumsy knight, was fleeing from a fire breathing dragon
     when he suddenly noticed the castle wall before him. Fortunately for
     clumsy Chuck, Merlin whisked him into the castle as the dragon crashed
     snout-first into the wall. Chuck's position was   10t   kilometers from
     the dragon's lair   t   hours after the dragon began chasing him. Earlier,
     clumsy Chuck had been quite still at the dragon's lair while trying to
     figure out how to get a dragon-skin coat. Let   y(t)   be Chuck's position
     and   v(t)   be his speed. Merlin saved Chuck at $t_0 = 1/2$.

     (a)   Find $\lim_{t \to 0} y(t)$, $\lim_{t \to 0} v(t)$, $\lim_{t \to 1/4} y(t)$,   and $\lim_{t \to 1/4} v(t)$.

     (b)   If   v(t)   suddenly became zero at $t_0 = 1/2$,   does $\lim_{t \to 1/2} v(t)$
           exist? Explain.

ANSWERS TO PREREQUISITE QUIZ

1.   (a) $(x^2 + y^2)(x + y)(x - y)$

     (b) $(x - 2)(x^2 + 2x + 4)$

     (c) $4(x + 2)^2$

     (d) $(3x + 2)(x + 2)$

2.   (a)   (b)

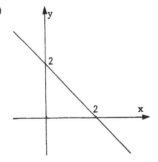   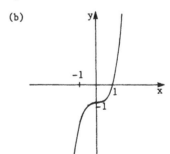

ANSWERS TO SECTION QUIZ

1. (a) 1

    (b) 0

    (c) Does not exist

2. (a) 1/12

    (b) 0

    (c) 2/5

    (d) 1/2

3. 60

4. (a) $\lim_{t \to 0} y(t) = 0$ kilometers; $\lim_{t \to 0} v(t)$ does not exist because the speed

        "jumps" from 0 to 10 km/hr; $\lim_{t \to 1/4} y(t) = 5/2$ kilometers ;

        $\lim_{t \to 1/4} v(t) = 10$ km/hr.

    (b) No, because the speed approaches 0 km/hr on one side of t_0

        and it approaches 10 km/hr on the other side.

1.3 The Derivative as a Limit and the Leibniz Notation

PREREQUISITES

1. Recall the $\Delta y/\Delta x$ method for computing derivatives (Section 1.1).

2. Recall formulas used to differentiate quadratic and linear functions (Section 1.1).

3. Recall the laws of limits (Section 1.2).

PREREQUISITE QUIZ

1. Use the $\Delta y/\Delta x$ method to differentiate $5x^2 + 3$.

2. (a) What is $\Delta y/\Delta x$ for $f(x) = 2x + 15$?

    (b) Use the result from part (a) to find $f'(3)$.

3. Differentiate the following functions:

    (a) $2 - 4x^2$

    (b) $5x$

    (c) $x^2 - 2x + 4$

    (d) $ax^2 + bx + c$, where a,b, and c are constants.

4. Compute the following limits:

    (a) $\lim_{x \to 2}(x^2 - x)$

    (b) $\lim_{x \to 3}[(x^3 - 27)/(x - 3)]$

GOALS

1. Be able to state and to use the definition of the derivative.

2. Be able to interchange the Leibniz notation, dy/dx, with other notations for the derivative, such as $f'(x)$.

STUDY HINTS

1. Derivative defined. It is defined as $\lim_{\Delta x \to 0} (\Delta y / \Delta x)$, where Δy is

    $f(x + \Delta x) - f(x)$. Previously, you saw the less precise form in

    Section 1.1 in which Δx was "replaced" by zero. Now, we take the

    limit of the difference quotient.

2. Continuity and differentiability. One implies the other; however, the

    reverse is not always true. What implication direction is true?

    (See Example 5.)

3. Nondifferentiability. A function is differentiable only if its graph

    is "smooth." A "pointed" function like the one in Fig. 1.3.2 is not

    differentiable at the "point."

4. Leibniz notation. dy/dx is just another notation for the derivative.

    It is not a fraction, so do not treat it as numerator and denominator;

    however, in many derivations, it acts like a fraction.

5. More notation. $(dy/dx)\big|_{x_0}$ means to evaluate the derivative at x_0 .

SOLUTIONS TO EVERY OTHER ODD EXERCISE

1. The definition of the derivative is $f'(x) = \lim_{\Delta x \to 0} (\Delta y / \Delta x) =$

    $\lim_{\Delta x \to 0} \{[f(x + \Delta x) - f(x)] / \Delta x\}$. Here, $\Delta y = [(x + \Delta x)^2 + (x + \Delta x)] -$

    $(x^2 + x) = [(x^2 + 2x\Delta x + (\Delta x)^2 + (x + \Delta x)] - (x^2 + x) = 2x\Delta x + (\Delta x)^2 +$

    Δx ; $\Delta y / \Delta x = 2x + \Delta x + 1$. Thus, $\lim_{\Delta x \to 0} (\Delta y / \Delta x) = 2x + 1$.

5. The definition of the derivative is $f'(x) = \lim_{\Delta x \to 0} (\Delta y / \Delta x) =$

    $\lim_{\Delta x \to 0} \{[f(x + \Delta x) - f(x)] / \Delta x\}$. Here, $\Delta y = 3/(x + \Delta x) - 3/x =$

    $[3x - 3(x + \Delta x)] / x(x + \Delta x) = -3\Delta x / x(x + \Delta x)$; $\Delta y / \Delta x = -3/x(x + \Delta x)$.

    Thus, $\lim_{\Delta x \to 0} (\Delta y / \Delta x) = -3/x^2$.

9. The definition of the derivative is $f'(x) = \displaystyle\lim_{\Delta x \to 0} (\Delta y / \Delta x) =$
    $\displaystyle\lim_{\Delta x \to 0} \{[f(x + \Delta x) - f(x)]/\Delta x\}$. Here, $\Delta y = 2\sqrt{x + \Delta x} - 2\sqrt{x} =$
    $2(\sqrt{x + \Delta x} - \sqrt{x})$. Multiplying top and bottom by $\sqrt{x + \Delta x} + \sqrt{x}$
    yields $\Delta y = 2\Delta x/(\sqrt{x + \Delta x} + \sqrt{x})$; $\Delta y / \Delta x = 2/(\sqrt{x + \Delta x} + \sqrt{x})$.
    Thus, $\displaystyle\lim_{\Delta x \to 0} (\Delta y / \Delta x) = 2/2\sqrt{x} = 1/\sqrt{x}$.

13. The difference quotient at $x_0 = 0$ is $[(1 + |0 + \Delta x|) -$
    $(1 + |0|)]/\Delta x = |\Delta x|/\Delta x$, which has no limit according to Example 5.
    Thus, the derivative at $x_0 = 0$ does not exist. On the other hand,
    $f(0) = 1$ and $\displaystyle\lim_{x \to 0} (1 + |x|) = \displaystyle\lim_{x \to 0} 1 + \displaystyle\lim_{x \to 0} |x| = 1$, as shown in
    Example 5. Since $\displaystyle\lim_{x \to 0} f(x) = f(0)$, $f(x)$ is continuous at $x_0 = 0$.

17. dy/dx is a new notation for the derivative. $\Delta y = [3(x + \Delta x)^3 +$
    $(x + \Delta x)] - (3x^3 + x) = [3x^3 + 9x^2\Delta x + 9x(\Delta x)^2 + 3(\Delta x)^3 + x + \Delta x] -$
    $(3x^3 + x) = 9x^2\Delta x + 9x(\Delta x)^2 + 3(\Delta x)^3 + \Delta x$, so $\Delta y/\Delta x = 9x^2 + 9x\Delta x +$
    $3(\Delta x)^2 + 1$. Thus, $\displaystyle\lim_{\Delta x \to 0} (\Delta y / \Delta x) = dy/dx = 9x^2 + 1$.

21. The velocity is given by $f'(1)$, which is the limit of $[f(1 + \Delta t) -$
    $f(1)]/\Delta t$ as Δt approaches 0 . $[f(1 + \Delta t) - f(1)]/\Delta t = [5(1 + \Delta t)^3 -$
    $5(1)^3]/\Delta t = [5(1 + 3\Delta t + 3(\Delta t)^2 + (\Delta t)^3) - 5]/\Delta t = 15 + 15\Delta t + 5(\Delta t)^2$.
    The limit is 15 , which is the velocity.

25. Using the result of Exercise 17, we have $(d/dx)(3x^3 + x)|_{x=1} =$
    $(9x^2 + 1)|_{x=1} = 10$.

29. The derivative is $\displaystyle\lim_{\Delta x \to 0} \{[1/(x + \Delta x)^2 - 1/x^2]/\Delta x\} = \displaystyle\lim_{\Delta x \to 0} \{[x^2 -$
    $(x^2 + 2x\Delta x + (\Delta x)^2)]/\Delta x(x + \Delta x)^2 x^2\} = \displaystyle\lim_{\Delta x \to 0} [-(2x + \Delta x)/(x + \Delta x)^2 x^2] =$
    $-2x/x^4 = -2/x^3$.

33.

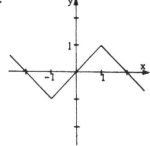

There is no unique answer. Following the result of Example 5, we wish to find a function with two "points" as shown in the figure. Such a function is:

$$f(x) = \begin{cases} -x - 2 & x \leq -1 \\ x & -1 \leq x \leq 1 \\ -x + 2 & x \geq 1 \end{cases}.$$

Two other possible answers are given in the back of the text.

SECTION QUIZ

1. State the definition of the derivative.

2. True or false: $|x|$ is not differentiable anywhere. Explain.

3. Evaluate $[(d/dx)(3x^2/4 - 5x)]\big|_3$.

4. Tough Tommy was enjoying his camping trip until the Thing woke him up.
    Upon seeing the Thing, tough Tommy almost jumped out of his pajamas.
    His position during the first hour is given by $x = 7t - t^2 - t^3$,
    where x is in miles and t is in hours.

    (a) Use the definition of the derivative to find his instantaneous
         speed, dx/dt .

    (b) Apply the definition of the derivative again to find his instan-
         taneous acceleration, d^2x/dt^2 .

ANSWERS TO PREREQUISITE QUIZ

1. $\Delta y/\Delta x = 10x + 5\Delta x$, so $f'(x) = 10x$.

2. (a) 2

    (b) 2

3. (a) -8x

    (b) 5

    (c) 2x - 2

    (d) 2ax + b

4. (a) 2

    (b) 27

ANSWERS TO SECTION QUIZ

1. The definition is $f'(x) = \lim_{\Delta x \to 0} \{[f(x + \Delta x) - f(x)]/\Delta x\}$.

2. False; it is not differentiable only at $x = 0$. Elsewhere, $f'(x) =$
    -1 if $x < 0$ and $f'(x) = 1$ if $x > 0$.

3. -1/2

4. (a) $7 - 2t - 3t^2$ miles/hour

    (b) $-2 - 6t$ miles/hour^2

1.4   Differentiating Polynomials

PREREQUISITES

1.   Recall how to compute derivatives using the limit method (Section 1.3).

PREREQUISITE QUIZ

1.   Describe the limit method for computing derivatives.

GOALS

1.   Be able to combine the power rule, the sum rule, and the constant
     multiple rule to differentiate polynomials.

STUDY HINTS

1.   **Basic differentiation rules**. Memorize $(d/dx)x^n = nx^{n-1}$, $(kf)'(x) =$
     $kf'(x)$, and $(f + g)'(x) = f'(x) + g'(x)$.

2.   **Polynomial rule**. This rule incorporates the three basic rules intro-
     duced in this section. Don't memorize it, but you should be able to
     use the three basic rules to differentiate any polynomial.

SOLUTIONS TO EVERY OTHER ODD EXERCISE

1.   By the power rule, $(d/dx)x^{10} = 10x^9$.

5.   By the power and constant mulitple rules, $(d/dx)(-5x^4) = 5(4x^3) = -20x^3$.

9.   By the power and constant mulitple rules, $(d/dx)(3\sqrt{x}) = (3(1/2)(x^{-1/2}) =$
     $3/2\sqrt{x}$.

13. $f(x) + g(x) = 3x^2 + x + 13$, so $(f + g)'(x) = 6x + 1$. On the other
     hand, $f'(x) = g'(x) = (6x) + (1)$.

17. By the sum rule, the derivative is $f'(x) - [2g(x)]'$. By the constant multiple rule, this is $f'(x) - 2g'(x)$.

21. By the polynomial rule, the derivative is $5t^4 + 12t + 8$.

25. By the polynomial rule, $g'(s) = 13s^{12} + 96s^7 - 21s^6/8 + 4s^3 + 4s^2$.

29. By the polynomial rule, $g'(h) = 80h^9 + 9h^8 - 113h$.

33. Multiplying out, $(t^3 - 17t + 9)(3t^5 - t^2 - 1) = 3t^8 - 51t^6 + 26t^5 + 16t^3 - 9t^2 + 17t - 9)$. Thus, the polynomial rule yields $f'(t) = 24t^7 - 306t^5 + 130t^4 + 48t^2 - 18t + 17$.

37. Multiplication gives $h(t) = t^7 - t^5 + 9t^3 - 9t$, so the polynomial rule gives $h'(t) = 7t^6 - 5t^4 + 27t^2 - 9$.

41. By the sum rule and Example 4, Section 1.3, $f'(x) = 2x - 1/2\sqrt{x}$.

45. Multiplication gives $f(x) = 1 - x$, so $f'(x) = -1$.

49. The slope of the tangent line is $(d/dx)(x^4 - x^2 + 3x)|_1 = (4x^3 - 2x + 3)|_1 = 5$.

53. Starting with $(d/dx)x^n = nx^{n-1}$, we guess that the desired function is x^{n+1} . Differentiating this gives $(n + 1)x^n$. So we now guess $x^{n+1}/(n+1)$, whose derivative is x^n . Noting that the derivative of a constant is zero, our desired function is $x^{n+1}/(n+1) + C$, where C is any constant.

57. Recall from basic geometry that $V(r) = 4\pi r^3/3$ and that the surface area is $A(r) = 4\pi r^2$. (These formulas are on the inside front cover of the text.) By the constant multiple and power rules, $V'(r) = (4\pi/3)(3r^2) = 4\pi r^2 = A(r)$.

SECTION QUIZ

1.   In each case, compute dy/dx.

(a) $y = x^{91} - x^{53} + x$

(b) $y = -4x^6 + 8x^3$

(c) $y = 2x^5(x^2 + 1)$

(d) $y = a^2 + a$,   a   is constant

2.   Find the equation of the line perpendicular to the curve $y = x^4/4 -$
$2x^3 + 3x - 1$ at $x_0 = 2$.

3.   (a) Compute du/dy for $u = y^5 - y^3/2 + y$.

(b) Compute g'(1) for $g(t) = 5t^5 + 4t^4 + 3t^3 + 2t^2 + t + 1$.

4.   Your hillbilly relatives are visiting the city for the first time. At
the local department store, your young cousin tries to run up the down
escalator. Horizontally, the escalator moves 30 feet in 10 seconds.
Since your cousin tires as he runs up, his position on the nonmoving
escalator with respect to the ground is given by $x = t^2 - t^3/6 - t^4/40$
for $0 \leqslant t \leqslant 3$. Answer the following questions with respect to the
ground.

(a) Compute the speed of the escalator.

(b) Compute the speed of your hillbilly cousin on the stationary escalator.

(c) Use your answers in parts (a) and (b) to compute his speed on the
moving escalator

ANSWERS TO PREREQUISITE QUIZ

1.   Compute the difference quotient $\Delta y/\Delta x = [f(x + \Delta x) - f(x)]/\Delta x$.
Simplify as much as possible and then take the limit as Δx approaches 0 .

ANSWERS TO SECTION QUIZ

1. (a) $91x^{90} - 53x^{52} + 1$

   (b) $-24x^5 + 24x^2$

   (c) $14x^6 + 10x^4$

   (d) 0

2. $13y - x + 93 = 0$

3. (a) $5y^4 - 3y^2/2 + 1$

   (b) 55

4. (a) 3 feet/second .

   (b) $(2t - t^2/2 - t^3/10)$ feet/second .

   (c) $(2t - t^2/2 - t^3/10 - 3)$ feet/second .

1.5   Products and Quotients

PREREQUISITES

1.   Recall how to compute limits (Section 1.3).

2.   Recall how to differentiate polynomials (Section 1.4).

PREREQUISITE QUIZ

1.   What is $\lim_{x \to 0} [(x + 5)(x^2 + 3x)/x]$?

2.   Differentiate the following functions:

   (a) $f(x) = x^{50} + 30x^{25} + x^4 + 50$.

   (b) $f(x) = 3x^{70} - 2x^{48} + 5(10)^6$.

GOALS

1.   Be able to state and apply the product rule.

2.   Be able to state and apply the quotient rule.

STUDY HINTS

1.   <u>Product rule.</u>   Memorize the formula $(fg)'(x) = f(x)g'(x) + f'(x)g(x)$.
     You will probably not be required to reproduce its derivation. You're
     wasting time if you're using this rule when one of f or g is con-
     stant; use the constant multiple rule.

2.   <u>Quotient rule.</u>   Memorize the formula $(f/g)'(x) = [f'(x)g(x) - f(x)g'(x)]/$
     $[g(x)]^2$.   The memory aid preceding Example 5 may be useful to you. As
     with the product rule, don't be too concerned with the derivation of the
     quotient rule.

3.   <u>Reciprocal rule.</u>   An alternative to learning this rule is to apply the
     quitient rule with $f(x) = 1$;   however, you will find that the reciprocal
     rule is faster to use.

4. Integer power rule. This is just an extension of the rule presented
    in Section 1.4. The rule you learned earlier now holds for any integer.
    CAUTION: A very common mistake is to add 1 to the exponent rather
    than subtracting 1. Thus, many students equate $d(x^{-2})/dx$ with $-2x^{-1}$
    rather than the correct answer, $-2x^{-3}$. Be careful!

5. Differentiation rules. At this point in your studies, you should know
    the rules for differentiating a power, a sum, a constant multiple, a
    product, and a quotient. All of the others are special cases of these
    rules and need not be memorized.

SOLUTIONS TO EVERY OTHER ODD EXERCISE

1. By the product rule, $(d/dx)[(x^2 + 2)(x + 8)] = (x^2 + 2)(1) + (2x)(x + 8)$:
    $3x^2 + 16x + 2$. On the other hand, $(x^2 + 2)(x + 8) = x^3 + 8x^2 + 2x + 16$
    so the derivative is $3x^2 + 16x + 2$.

5. By the product rule, $(d/dx)[(x^2 + 2x + 1)(x - 1)] = (x^2 + 2x + 1)(1) +$
    $(2x + 2)(x - 1) = 3x^2 + 2x - 1$. On the other hand, $(x^2 + 2x + 1)(x - 1)$
    $x^3 + x^2 - x - 1$, so the derivative is $3x^2 + 2x - 1$.

9. By the product rule, $(d/dx)[(x - 1)(x^2 + x + 1)] = (x - 1)(2x + 1) +$
    $(1)(x^2 + x + 1) = 3x^2$. On the other hand, $(x - 1)(x^2 + x + 1) =$
    $x^3 - 1$, so the derivative is $3x^2$.

13. $d(x^{5/2})/dx = (d/dx)(x^2 \cdot \sqrt{x}) = 2x\sqrt{x} + x^2/2\sqrt{x} = 5x^{3/2}/2$.

17. By the quotient rule, $(d/dx)[(x - 2)/(x^2 + 3)] = [(1)(x^2 + 3) -$
    $(x - 2)(2x)]/(x^2 + 3)^2 = (-x^2 + 4x + 3)/(x^2 + 3)^2$.

21. By the quotient rule, $(d/dx)[(x^2 + 2)/(x^2 - 2)] = [(2x)(x^2 - 2) -$
    $(x^2 + 2)(2x)]/(x^2 - 2)^2 = -8x/(x^2 - 2)^2$.

25. By the quotient rule, $(d/dr)[\,(r^2 + 2)/r^8\,] = [\,(2r)(r^8) - (r^2 + 2)(8r^7)\,]/$
    $r^{16} = (-6r^9 - 16r^7)/r^{16} = (-6r^2 - 16)/r^9$.

29. We expand, first, to get $((x^2 + 1)^2 + 1)/((x^2 + 1)^2 - 1) = (x^4 + 2x^2 + 2)/$
    $(x^4 + 2x^2)$. By the quotient rule, the derivative is
    $[\,(4x^3 + 4x)(x^4 + 2x^2) - (x^4 + 2x^2 + 2)(4x^3 + 4x)\,]/(x^4 + 2x^2)^2 =$
    $(-8x^3 - 8x)/(x^4 + 2x^2)^2$.

33. By the reciprocal rule, $(d/dx)[\,1/(x + 1)^2\,] = (d/dx)[\,1/(x^2 + 2x + 1)\,] =$
    $-(2x + 2)/(x^2 + 2x + 1)^2 = -2(x + 1)/(x + 1)^4 = -2/(x + 1)^3$.

37. Combining the reciprocal rule with other derivative rules, we get
    $(d/dy)[\,4(y + 1)^2 - 2(y + 1) - 1/(y + 1)\,] = (d/dy)[\,4(y^2 + 2y + 1)\,] -$
    $2 + 1/(y + 1)^2 = 8y + 6 + (y + 1)^{-2}$.

41. It is easiest to mulitply first to get $(3\sqrt{x} + 1)x^2 = 3x^{5/2} + x^2$.
    Then, the derivative is $15x^{3/2}/2 + 2x$. On the other hand, the product
    rule may be used to obtain $(3/2\sqrt{x})x^2 + (3\sqrt{x} + 1)(2x) = 15x^{3/2}/2 + 2x$.
    Both methods give the same result.

45. By the reciprocal rule, $(d/dx)[\,\sqrt{2}/(1 + 3\sqrt{x})\,] = -\sqrt{2}(3/2\sqrt{x})/(1 + 3\sqrt{x})^2 =$
    $-3/\sqrt{2x}(1 + 3\sqrt{x})^2$.

49. By the reciprocal rule, $f'(x) = -(1/2\sqrt{x})/(\sqrt{x})^2 = -1/2x^{3/2}$.
    Thus, the slope of the tangent line at $x = 2$ is $(-1/2x^{3/2})|_2 =$
    $-1/4\sqrt{2} = -\sqrt{2}/8$.

53. By the product and sum rules, the derivative is $[\,f(x) + xf'(x)\,] +$
    $g'(x) = (4x^5 - 13x) + x(20x^4 - 13) + (3x^2 + 2) = 24x^5 + 3x^2 - 26x + 2$.

57. We apply the reciprocal rule to get $(d/dx)(1/P(x)) = -P'(x)/[\,P(x)\,]^2$.
    Let $P(x) = ax^2 + bx + c$, so $-P'(x) = -2ax - b = 0$. The only solu-
    tion is $x = -b/2a$ if $P(-b/2a) \neq 0$. $(d/dx)(1/P(x))$ is never zero
    when $P(x)$ is a perfect square such as x^2 or $x^2 + 2x + 1$.

SECTION QUIZ

1. By the quotient rule, $(d/dx)(2/x^3) = [(0)(x^3) - (2)(3x^2)]/(x^3)^2 =$ $-6x^2/x^6 = -6x^{-4}$. On the other hand, $2/x^3 = 2x^{-3}$, so by the integer power rule, $(d/dx)(2x^{-3}) = (2)(-3x^{-2}) = -6x^{-2}$. Why aren't the two derivatives equal?

2. Suppose $(d/dy)\omega = \Omega$, $(d/dy)\gamma = \Gamma$, and $(d/dy)\delta = \Delta$. What is $(d/dy)[(\omega\gamma - \gamma\delta)/\omega\delta]$?

3. Differentiate $(x^2 + 2x + \sqrt{x})(x^2 + 3)/(x^2 - 2)(x + \sqrt{x})$.

4. The mad scientist, Liver Louie, enjoys eating chicken livers. After performing numerous genetic experiments, he can produce chickens with $2x - 1$ livers in the xth generation. Also, he has increased the size of each liver so that each weighs $x^2 + x + 1$ ounces.

    (a) Use the product rule to determine how fast liver weight increases in each chicken during the xth generation.

    (b) If total revenue, R , from one chicken's liver is $x^2 + 3x + 1$ dollars, what is the inflation rate for chicken livers in the xth generation, i.e., what is $(d/dx)(R/Q)$ where Q is one chicken's liver weight?

ANSWERS TO PREREQUISITE QUIZ

1. 15

2. (a) $50x^{49} + 750x^{24} + 4x^3$

    (b) $210x^{69} + 96x^{47}$

ANSWERS TO SECTION QUIZ

1. By the integer power rule, the derivative of x^{-3} is $-3x^{-4}$, not
    $-3x^{-2}$.

2. $[(\Omega\gamma + \omega\Delta - \Gamma\delta - \gamma\Delta)\omega\delta - (\omega\gamma - \gamma\delta)(\Omega\delta + \omega\Delta)]/\omega^2\delta^2$

3. $\{[(2x + 2 + 1/2\sqrt{x})(x^2 + 3) + (x^2 + 2x + \sqrt{x})(2x)](x^2 - 2)(x + \sqrt{x}) - (x^2 + 2x + \sqrt{x})(x^2 + 3)[(2x)(x + \sqrt{x}) + (x^2 - 2)(1 + 1/2\sqrt{x})]\}/(x^2 - 2)^2(x + \sqrt{x})^2$

4. (a) $6x^2 + 2x + 1$

    (b) $[(2x + 3)(2x - 1)(x^2 + x + 1) - (6x^2 + 2x + 1)(x^2 + 3x + 1)]/$
         $(2x - 1)^2(x^2 + x + 1)^2$

1.6   The Linear Approximation and Tangent Lines.

PREREQUISITES

1.   Recall the point-slope form of the line (Section R.4).

2.   Recall the relationship between $f'(x)$ and $\Delta y/\Delta x$, i.e., recall the
     definition of the derivative (Section 1.3).

PREREQUISITE QUIZ

1.   What is the equation of a line with slope m and passing through the
     point (x_0, y_0) ?

2.   How is $f'(x)$ related to $\Delta y/\Delta x$?

GOALS

1.   Be able to write the equation of the tangent line to a curve at a
     given point.

2.   Be able to find an approximation by using derivatives.

STUDY HINTS

1.   Tangent lines. The equation doesn't need to be memorized: if you
     know $y = y_0 + m(x - x_0)$, just substitute $y_0 = f(x_0)$ and $m = f'(x_0)$.

2.   Linear approximation. You should either memorize or learn to derive
     the formula. If you remember that $[f(x_0 + \Delta x) - f(x_0)]/\Delta x \approx f'(x_0)$
     from Section 1.1, you can easily derive the formula by multiplying
     through by Δx and then adding $f(x_0)$ to both sides. Note that Δx
     may be negative. Using a calculator defeats the purpose of learning what
     the linear approximation is all about.

3.   <u>Relationship between tangency and the approximation.</u>   Notice that the

linear approximation was derived from $\Delta y/\Delta x \approx f'(x_0)$.   Also notice

that the tangent line may be rewritten as $m = (y - y_0)/(x - x_0)$ or

$f'(x_0) = \Delta y/\Delta x$.   Thus, one basic formula gives you the concepts of

this section.

SOLUTIONS TO EVERY OTHER ODD EXERCISE

1.   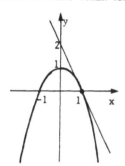   The tangent line to $y = f(x)$ is $y = f(x_0) +$

$f'(x_0)(x - x_0)$.   For $y = 1 - x^2$, $f'(x) =$

$-2x$,   so $y = 0 + (-2)(x - 1) = -2x + 2$.

5.   The tangent line to $y = f(x)$ is $y = f(x_0) + f'(x_0)(x - x_0)$.   Com-

bining the product and quotient rules, we have $f'(x) = (2x)[3x/(x + 2)] +$

$(x^2 - 7)[3(x + 2) - 3x(1)]/(x + 2)^2$,   so $f'(x_0) = 0[0/2] + (-7)[(3)(2) -$

$(0)(1)]/(2)^2 = -21/2$.   Also, $f(x_0) = -7(0/2) = 0$;   therefore, the

tangent line is $y = (-21/2)x$.

9.   By the quotient rule, $f'(x) = [(1)(x + 1) - (x)(1)]/(x + 1)^2 =$

$1/(x + 1)^2$ and $f'(x_0) = 1/4$.   Since $f(x_0) = 1/2$,   the tangent line

is $y = 1/2 + (1/4)(x - 1)$.   To find the x-intercept, set $y = 0$.

This gives $x = -1$ as the crossing point.

13. Use the approximation $f(x_0 + \Delta x) \approx f(x_0) + f'(x_0)\Delta x$.   Let $f(x) = x^2$,

so $f'(x) = 2x$.   Thus, $f(2.02) = f(2 + 0.02) \approx f(2) + f'(2)(0.02) = 4 +$

$4(0.02) = 4.08$.   The exact value is   4.0804 .

17. Use the approximation $f(x_0 + \Delta x) \approx f(x_0) + f'(x_0)\Delta x$. Let $f(x) = \sqrt{x}$, so $f'(x) = 1/2\sqrt{x}$. Thus, $f(16.016) = f(16 + 0.016) \approx f(16) + f'(16)(0.016) = 4 + (1/8)(0.016) = 4.002$. The exact value is approximately 4.001999 .

21. Using the approximation $f(x_0 + \Delta x) \approx f(x_0) + f'(x_0)\Delta x$, we let $f(x) = x^4$, so $f'(x) = 4x^3$. Thus, $f(2.94) = f(3 - 0.06) \approx f(3) + f'(3)(-0.06) = 81 + 108(-0.06) = 74.52$.

25. Let $A(r) = \pi r^2$, so $A'(r) = 2\pi r$. $A(3.04) = A(3 + 0.04) \approx A(3) + A'(3)(0.04)$. Therefore, the approximate increase in area is $A'(3)(0.04) = 6\pi(0.04) = 0.24\pi$.

29. We use $f(x_0 + \Delta x) \approx f(x_0) + f'(x_0)\Delta x$ with $x_0 = 3$ and $\Delta x = 0.023$. Thus, $f(x_0) = (12)(5) = 60$ and $f'(x) = (2x)(x + 2) + (x^2 + 3)(1)$, and so $f'(x_0) = (6)(5) + (12)(1) = 42$. Therefore, $f(3.023) \approx 60 + 42(0.023) = 60.966$.

33. The tangent line has the equation $y = f(1) + f'(1)(x - 1)$. Here, $f'(x) = 8x^7 + 4x$, so $f'(1) = 12$. Since $f(1) = 4$, the equation is $y = 4 + 12(x - 1)$ or $y = 12x - 8$.

37. We use $f(x_0 + \Delta x) \approx f(x_0) + f'(x_0)\Delta x$ with $x_0 = 1$ and $\Delta x = -0.0003$. Here, $f'(s) = 4s^3 - 15s^2 + 3$, so $f'(1) = 4 - 15 + 3 = -8$. Also, $f(1) = 1 - 5 + 3 - 4 = -5$, so $f(0.9997) \approx -5 + (-8)(-0.0003) = -4.9976$.

41. For each case, we use $h(t_0 + \Delta t) \approx h(t_0) + f'(t_0)\Delta t$, where $h'(t) = -8t + 7$.

$h(3.001) = h(3 + 0.001) \approx h(3) + h'(3)(0.001) = -14.25 + (-17)(0.001)$
$= -14.267$.

$h(1.97) = h(2 - 0.03) \approx h(2) + h'(2)(-0.03) = -1.25 + (-9)(-0.03) = -0.98$.

$h(4.03) = h(4 + 0.03) \approx h(4) + h'(4)(0.03) = -35.25 + (-25)(0.03) = -36.00$

45. (a) $f'(x) = 4x^3$ and the linear approximation to $f(x)$ near $x = 2$

is $f(2) + f'(2)\Delta x = (2)^4 + 4(2)^3\Delta x = 16 + 32\Delta x$.

(b) In general, the linear approximation at $x_0 + \Delta x$ is $x_0^4 + 4x_0^3\Delta x$.

The actual value is $(x_0 + \Delta x)^4 = x_0^4 + 4x_0^3\Delta x + 6x_0^2(\Delta x)^2 + 4x_0(\Delta x)^3 +$

$(\Delta x)^4$. Thus, the difference between the linear approximation and

the actual value is $-[6x_0^2(\Delta x)^2 + 4x_0(\Delta x)^3 + (\Delta x)^4] =$

$-(\Delta x)^2[6x_0^2 + 4x_0\Delta x + (\Delta x)^2] = -(\Delta x)^2[(2x_0 + \Delta x)^2 + 2x_0^2]$, which

is $\leqslant 0$. Thus, the linear approximation is smaller than the actual

value.

(c) We want $6x_0^2(\Delta x)^2 + 4x_0(\Delta x)^3 + (\Delta x)^4 \leqslant 1.6$, where $x_0 = 2$, i.e.,

$24(\Delta x)^2 + 8(\Delta x)^3 + (\Delta x)^4 \leqslant 1.6$. Therefore, the interval is

$[1.730, 2.247]$.

49. The following chart includes a few examples to verify that $1 - x + x^2$

is indeed a better approximation of $1/(1 + x)$ than $1 - x$.

x	$1/(1 + x)$	$1 - x$	$1 - x + x^2$
0.1	0.9090909091	0.9	0.91
0.01	0.9900990099	0.99	0.9901
0.001	0.999000999	0.999	0.999001
0.0001	0.99990001...	0.9999	0.99990001

The above calculations demonstrate that an extra term will provide a

better approximation. By long division, the first three terms of

$3600/(60 + x)$ are $(60 - x + x^2/60)$ miles/hour .

SECTION QUIZ

1. Find an approximate value for $3(2.008)^2 - 5(2.008) + 3$.

2. A cube has edges 4.989 cm. long. Estimate its surface area.

3. Consider the graph of the equation $y = x^3 + 2x - 1$

(a) What is the tangent line at $x = 2$.

(b) Where does the tangent line at $x = -1$ intersect the tangent line in part (a)?

4. Cathy Cuisine enjoys dining at the city's finest gourmet restaurants. However, she never carries cash and somebody had just stolen all of her National Rapids money orders. The furious restaurateur questions her, "What will you do? What will you do? Do you want to be dumped in the slammer or wash $x^3/100 + x + 65$ dishes for each x dollars you owe us?"

(a) Use the $\Delta y/\Delta x$ method to compute dy/dx for $y = x^3/100 + x + 65$.

(b) Suppose Cathy's dinner cost $35 . Use the linear approximation with $x_0 = 30$ to estimate how many dishes she must wash.

(c) Use $x_0 = 40$ to find another approximation.

(d) If she washes 100 dishes/hour , about how many hours must she wash dishes?

ANSWERS TO PREREQUISITE QUIZ

1. $y = y_0 + m(x - x_0)$

2. $f'(x) = \lim_{\Delta x \to 0} (\Delta y/\Delta x)$

ANSWERS TO SECTION QUIZ

1. 5.056

2. The surface area is $6x^2$, so $6(4.989)^2 \approx 149.34$ cm^2 .

3. (a) $y = 14x - 17$

(b) $(2 , 11)$

4.   (a) $3x^2/100 + 1$

     (b)   505

     (c)   500

     (d)   About 5 to 5.05 hours

1.R   Review Exercises for Chapter 1

SOLUTIONS TO EVERY OTHER ODD EXERCISE

1.   By the quadratic function rule, $f'(x) = 2x$.

5.   By the quadratic function rule with $a = 0$, $f'(x) = 2$.

9.   By the polynomial rule, $f'(x) = -50x^4 + 24x^2$.

13.   Combining the sum rule and the rule for square roots, $f'(x) = 9x^2 - 1/\sqrt{x}$.

17.   By the quotient rule, $f'(x) = [2x(x^2 -1) - (x^2 + 1)(2x)]/(x^2 - 1)^2 =$

   $-4x/(x^2 - 1)^2$.

21.   First, expand $(s + 1)^2$ and then, by the product rule, $(d/ds)[(s + 1)^2 \cdot$

   $(\sqrt{s} + 2)] = (d/ds)[(s^2 + 2s + 1)(\sqrt{s} + 2) = (2s + 2)(\sqrt{s} + 2) +$

   $(s^2 + 2s + 1)(1/2\sqrt{s}) = (5s^2 + 8s^{3/2} + 6s + 8\sqrt{s} + 1)/2\sqrt{s}$.

25.   By the reciprocal rule, $(d/dt)(3t^2 + 2)^{-1} = -6t/(3t^2 + 2)^2$.

29.   Multiplying and using the reciprocal rule, $(d/dx)[1/\sqrt{x}(\sqrt{x} - 1)] =$

   $(d/dx)[1/(x - \sqrt{x})] = (1 - 1/2\sqrt{x})/(x - \sqrt{x})^2 = -(2\sqrt{x} - 1)/2\sqrt{x}(x - \sqrt{x})^2$.

33.   Factor out $(x - 1)$ to get $\lim\limits_{x \to 1} [(x^3 - 1)/(x - 1)] =$

   $\lim\limits_{x \to 1}[(x - 1)(x^2 + x + 1)/(x - 1)] = \lim\limits_{x \to 1}(x^2 + x + 1) = 3$.

37.   By the continuity of rational functions, $\lim\limits_{x \to 3}[(3x^2 + 2x)/x] =$

   $(27 + 6)/3 = 11$.

41.   We recognize the limit as the definition of the derivative, so the limit

   is $f'(x) = 4x^3 + 6x$.

45.   Divide by x^3/x^3 to get $\lim\limits_{x \to \infty} [(5/x + 4/x^3)/(5 + 9/x^2)]$. Now, using

   the fact that $\lim\limits_{x \to \infty} (1/x) = 0$, the limit becomes $0/5 = 0$.

49.   The definition of the derivative is $\lim\limits_{\Delta x \to 0} \{ [f(x_0 + \Delta x) - f(x_0)]/\Delta x \}$,

   so $f'(1) = \lim\limits_{\Delta x \to 0} \{ [f(1 + \Delta x) - f(1)]/\Delta x \} = \lim\limits_{\Delta x \to 0}\{[3(1 + \Delta x)^3 + 8(1 + \Delta x) -$

   $3(1)^3 - 8(1)]/\Delta x\} = \lim\limits_{\Delta x \to 0} \{ [9\Delta x + 9(\Delta x)^2 + 3(\Delta x)^3 + 8\Delta x]/\Delta x \} =$

   $\lim\limits_{\Delta x \to 0} [17 + 9\Delta x + 3(\Delta x)^2] = 17$.

53. The slope of the tangent line is $f'(x_0)$. For $y = x^3 - 8x^2$,
$f'(x) = 3x^2 - 16x$, and so the slope is $f'(1) = -13$.

57. The slope of the tangent line is $f'(x_0)$. For $y = 3x^4 - 10x^9$,
$f'(x) = 12x^3 - 90x^8$, and so the slope is $f'(0) = 0$.

61. The velocity is given by $f'(t_0)$. By the quotient rule, $f'(t) =$
$[(2t + 1/2\sqrt{t})(1 + \sqrt{t}) - (t^2 + \sqrt{t})(1/2\sqrt{t})]/(1 + \sqrt{t})^2$. At $t_0 = 1$,
the velocity is $f'(1) = [(5/2)(2) - 2(1/2)]/(2)^2 = 1$ meter per second.

65. The linear approximation is $f(x_0 + \Delta x) \approx f(x_0) + f'(x_0)\Delta x$. Let
$f(x) = \sqrt{x}$, so $f'(x) = 1/2\sqrt{x}$. Thus, $f(4.0001) = f(4 + 0.0001) \approx$
$f(4) + f'(4)(0.0001) = 2 + (1/4)(0.0001) = 2.000025$.

69. The linear approximation is $f(x_0 + \Delta x) \approx f(x_0) + f'(x_0)\Delta x$. $h'(s) =$
$12s^2 - 4s^3$, so $h(2.95) = h(3 - 0.05) \approx h(3) + h'(3)(-0.05) = 27 +$
$(0)(-0.05) = 27.00$.

73. The tangent line is given by $y = y_0 + f'(x_0)(x - x_0)$. By the quotient
rule, $f'(x) = [(3x^2)(x^3 + 11) - (x^3 - 7)(3x^2)]/(x^3 + 11)^2$ and $f'(2) =$
$[(12)(19) - (1)(12)]/(19)^2 = 216/361$, so the tangent line is $y =$
$(1/19) + (216/361)(x - 2) = (216x - 431)/361$.

77. By the sum rule, the derivative is $f'(x) + g'(x) = (4x - 5) + (3x/2 + 2) =$
$11x/2 - 3$, so at $x = 1$, the derivative is $5/2$.

81. The volume V is $(r)(r)(3r) = 3r^3$. The surface area is $A = 2(r)(r) +$
$4(r)(3r) = 14r^2$. Now, $dV/dr = 9r^2$, which is $9/14$ of A .

85. The slope of the tangent is $y' = 2x - 2$. The tangency point is
$(x , x^2 - 2x + 2)$, so the slope of the line passing through the origin
is $(x^2 - 2x + 2)/x = 2x - 2$. The equation reduces to $x^2 - 2 = 0$, so
x must be $+\sqrt{2}$. $x = -\sqrt{2}$ is not acceptable because we want a positive
slope. Therefore, the tangent line is $y = 2 - 2\sqrt{2} + 2 + (2\sqrt{2} - 2)(x - \sqrt{2}) =$
$(2\sqrt{2} - 2)x$.

89. (a) Let $f(x) = a_n x^n + a_{n-1} x^{n-1} + \ldots + a_2 x^2 + a_1 x + a_0$, $a_n \neq 0$,

and $g(x) b_m x^m + b_{m-1} x^{m-1} + \ldots + b_1 x + b_0$, $b_m \neq 0$. Then

$f(x)g(x) = a_n b_m x^{n+m}$ + lower degree terms. Since $a_n \neq 0$ and

$b_m \neq 0$, $a_n b_m \neq 0$ and so, $\deg[f(x)g(x)] = n + m$. Also,

$\deg[f(x)] + \deg[g(x)] = n + m$.

(b) Let $f(x) = (a_n x^n + a_{n-1} x^{n-1} + \ldots + a_0)/(b_k x^k + b_{k-1} x^{k-1} + \ldots + b_0)$,

$a_n \neq 0$ and $b_k \neq 0$, and $g(x) = (c_m x^m + c_{m-1} x^{m-1} + \ldots + c_0)/$

$d_\ell x^\ell + d_{\ell-1} x^{\ell-1} + \ldots + d_0)$, $c_m \neq 0$ and $d_\ell \neq 0$. Then $f(x)g(x) =$

$(a_n c_m x^{n+m}$ + lower degree terms)$/(b_k d_\ell x^{k+\ell}$ + lower degree terms). By

definition, $\deg[f(x)g(x)] = n + m - k - \ell$. Also, $\deg[f(x)] = n - k$

and $\deg[g(x)] = m - \ell$, so $\deg[f(x)] + \deg[g(x)] = n + m - k - \ell$.

(c) Define $f(x)$ as in part (b). Then $f'(x) = [(na_n x^{n-1} +$

$(n - 1)a_{n-1} x^{n-2} + \ldots + a_1) \cdot (b_k x^k + b_{k-1} x^{k-1} + \ldots + b_0) -$

$(a_n x^n + a_{n-1} x^{n-1} + \ldots + a_0) \cdot (kb_k x^{k-1} + (k - 1)b_{k-1} x^{k-2} + \ldots +$

$b_1)]/(b_k x^k + b_{k-1} x^{k-1} + \ldots + b_0)^2 = [(na_n b_k x^{n+k-1}$ + lower degree

terms) $- (ka_n b_k x^{n+k-1}$ + lower degree terms)$]/(b_k^2 x^{2k}$ + lower degree

terms) $= ((n - k)a_n b_k x^{n+k-1}$ + lower degree terms)$/(b_k^2 x^{2k}$ + lower

degree terms). Therefore, $\deg[f'(x)] = (n + k - 1) - (2k) = n - k - 1$

Also, $\deg[f(x)] - 1 = n - k - 1$. This assumes $n - k \neq 0$.

If $\deg[f(x)] = 0$, i.e., $n - k = 0$, then the numerator in $f'(x)$

has degree at most $n + k - 2$. The denominator has degree $2k$, so

$\deg[f'(x)]$ is at most $(n + k - 2) - 2k = n - k - 2 = -2$. This is

all that can be said, as the example $f(x) = (x^k + 1)/(x^k - 1)$ shows.

TEST FOR CHAPTER 1

1.   True or false:

   (a)   If $f'(x) = g'(x)$, then $f(x) = g(x)$.

   (b) $\lim_{x \to 1} [(x^2 - 1)/(x - 1)] = 0$.

   (c)   If $r = s^5 + 5s^4$, then $ds/dr = 5s^4 + 20s^3$.

   (d)   For any constant b , the derivative, with respect to x , of

         $bf(x)$ is $bf'(x)$.

   (e)   If $y = (x^3 + x^2)x$, then $(dy/dx)|_2$ can only be 44 .

2.   (a)   State the product rule.

   (b)   State the quotient rule.

3.   Differentiate $(x^3 + 3x + 2)(5x^4 - 3x^3 + x)/(x^2 + 5)$.   Do not simplify.

4.   (a)   State the definition of the derivative.

   (b)   Use the limit method to find the derivative of $x/(x + 3)$.

5.   Find $(dy/dx)|_1$ if $y = (3x^2 + x)/(x + 5)$.

6.   Let $h(z) = 5z^7 + 2z^5 - z^4 + 3z^3 - 11z + 6$.

   (a)   Find the tangent line at $z_0 = 1$.

   (b)   Find the line perpendicular to $h(z)$ at $z_0 = 1$.

   (c)   If h is in miles and z is in hours, what is the physical

         interpretation of dh/dz ?

   (d)   Find an approximation for $h(0.97)$.

7.   Find $(d/dt)[t^3/(t^2 + 1)]$.

8.   Differentiate the following functions in x :

   (a) $f(x) = (x^6 - x^4 + 2)(3x^3 + x^2 - 3x)$

   (b) $f(x) = (x^4 + x)/(2x^4 - x^3 + x - 1)$

   (c) $f(x) = (x^5 + x^4 + x^3)^{-1}$

9.   Let $f(x) = \begin{cases} x^3 & \text{if } x \leqslant 0 \\ x^2 & \text{if } x \geqslant 0 \end{cases}$.   Use limits to determine if $f'(0)$ exists.

10.   One day in prehistory, a caveman, whose cave entrance was located at

(1,3) of a parabolic hill described by $y = 4 - x^2$, teased a

pterydactyl flying overhead. The angry reptile flew down at a tangent

to the hill and barely skimmed the cave entrance.

(a)   What equation describes the flight path?

(b)   Where should a fire be built at $y = 0$ if the caveman wants roast

pterydactyl for dinner?

ANSWERS TO CHAPTER TEST

1.   (a)   False, $f(x)$ and $g(x)$ may differ by a constant.

(b)   False, the limit is 2 .

(c)   False, $dr/ds = 5s^4 + 20s^3$, not ds/dr .

(d)   True

(e)   True

2.   (a)   The derivative of $f(x)g(x)$ is $f'(x)g(x) + f(x)g'(x)$.

(b)   The derivative of $f(x)/g(x)$ is $[f'(x)g(x) - f(x)g'(x)]/[g(x)]^2$,

provided $g(x) \neq 0$.

3. $[((3x^2 + 3)(5x^4 - 3x^3 + x) + (x^3 + 3x + 2)(20x^3 - 9x^2 + 1))(x^2 + 5) -$

$(x^3 + 3x + 2)(5x^4 - 3x^3 + x)(2x)]/(x^2 + 5)^2$.

4.   (a) $f'(x) = \lim_{\Delta x \to 0} \{[f(x + \Delta x) - f(x)]/\Delta x\}$.

(b) $3/(x + 3)^2$

5.   19/18

6.   (a) $h(z) = 39z - 35$

(b) $h(z) = 4 - (z - 1)/39$

(c)   It is the velocity.

(d)   2.83

7. $(t^4 + 3t^2)/(t^2 + 1)^2$

8. (a) $(6x^5 - 4x^3)(3x^3 + x^2 - 3x) + (x^6 - x^4 + 2)(9x^2 + 2x - 3)$

   (b) $[(4x^3 + 1)(2x^4 - x^3 + x - 1) - (x^4 + x)(8x^3 - 3x^2 + 1)]/$

   $(2x^4 - x^3 + x - 1)^2$

   (c) $-(5x^2 + 4x + 3)/(x^4 + x^3 + x^2)^2$

9. $f'(0)$ exists and equals zero because $\displaystyle\lim_{\Delta x \to 0} \{[f(0 + \Delta x) - f(0)]/\Delta x\}$ is equal to zero independent of whether we let $\Delta x \to 0$ from the left or the right.

10. (a) $y = 5 - 2x$

    (b) $(5/2, 0)$

CHAPTER 2

RATES OF CHANGE AND THE CHAIN RULE

2.1 Rates of Change and the Second Derivative

PREREQUISITES

1. Recall how to differentiate polynomials, products, and quotients
    (Sections 1.4 and 1.5).

2. Recall how velocity and slopes are related to the derivative (Section 1.1).

PREREQUISITE QUIZ

1. Differentiate:

    (a) $5x^2 + x - 6$

    (b) $(x + 3)(x^2 - 2)$

    (c) $(x - 5)/(x^2 + 2)$

2. An object's position is given by $y = x^2 - 3x + 2$. What is its velocity
    at $x_0 = 5$?

3. Explain how slopes are related to the derivative.

GOALS

1. Be able to relate rates of change with the derivative.

2. Be able to compute the second derivative and understand its physical
    meaning.

STUDY HINTS

1. Rates of change. As previously discussed, an average rate of change is
    $\Delta y / \Delta x$ over a finite interval, Δx . An instantaneous rate of change
    is simply the derivative, $f'(x)$. The derivative can represent any
    rate of change, i.e., a change in one quantity due to a change in
    another. A linear or proportional change is a special rate of change
    where $\Delta y = k \Delta x$ for a constant k ; this implies $f'(x)$ is constant
    for all x .

2. Sign of derivative. Think about the many possible interpretations of
    the derivative (look at Fig. I.7 on page 4). The sign indicates the
    direction of the change.

3. Second derivatives. This is simply the derivative of the first deriva-
    tive function. If you are asked to evaluate $f''(x_0)$, do not substitute
    x_0 until you compute the second derivative; otherwise, your answer will
    be zero. Why? If x is time and $y = f(x)$ is position, the interpre-
    tation of $f''(x)$ is acceleration.

4. Leibniz notation. The second derivative is denoted $d^2 y / dx^2$. Note
    the positions of the "exponents;" this comes from writing $(d/dx)(d/dx)$.

5. Economic applications. Examples 11 and 12 introduce many new terms which
    are used in economics. In general, the word "marginal" implies a deriv-
    ative. Ask you instructor how much economic terminology you will be
    held responsible for.

SOLUTIONS TO EVERY OTHER ODD EXERCISE

1. If $r = \Delta y / \Delta x =$ slope and $y = y_0$ when $x = x_0$, then $y = y_0 + r(x - x_C$
    In this case, $y = 1 + 5(x - 4) = 5x - 19$.

5. The rate of change of price with respect to time is $\Delta P/\Delta t$ = (3.2 - 2)

    cents/(1984 - 1982) years = 0.6 cents/year. Now, $P = P_0 + (\Delta P/\Delta t)(t - t_0)$.

    To determine t when the price is 5 , we solve 5 = 2 + (0.6)(t - 1982)

    to get   t = 1987 . To find the price in 1991, we let   t = 1991 , and so

    P = 2 + (0.6)(1991 - 1982) = 2 + 5.4 = 7.4 cents/kilowatt-hour .

9. The average rate of change is $\Delta f(t)/\Delta t$. $\Delta f(t)$ = f(3/2) - f(1) =

    334-364 = -30 and Δt = 3/2 - 1 = 1/2 . Thus, the average rate of change

    is -30/(1/2) = -60 .

13. The area of the base with radius r is πr^2 and the height is r .

    Therefore, $V = \pi r^3/3$ and the rate of change of the volume with respect

    to the radius is $dV/dr = 3\pi r^2/3 = \pi r^2$.

17. The rate of change of H with respect to d is $H'(d)$ = 56 - 6d .

    Hence, at d = 0.5 , $H'(d)$ = 53 .

21. The rate of change of the volume with respect to the radius is $V'(r)$ =

    $4\pi r^2$, which is the surface area of the sphere.

25. By the product rule, the velocity is $(5t^4)(t + 2) + (t^5 + 1)(1)$ =

    $6t^5 + 10t^4 + 1$. Thus, the acceleration is the second derivative,

    $30t^4 + 40t^3$. At t = 0.1 , the velocity is 1.00106 and the accel-

    eration is 0.043 .

29. Applying the quotient rule twice, $(d^2/dx^2)[(x^2 + 1)/(x + 2)]$ =

    $(d/dx)\{[(2x)(x + 2) - (x^2 + 1)(1)]/(x + 2)^2\} = (d/dx)[(x^2 + 4x - 1)/$

    $(x^2 + 4x + 4)] = [(2x + 4)(x^2 + 4x + 4) - (x^2 + 4x - 1)(2x + 4)]/$

    $(x^2 + 4x + 4)^2 = (10x + 20)/(x + 2)^4 = 10/(x + 2)^3$.

33. $d^2f(x)/dx^2$ = d(2x)/dx = 2 .

37. $d^2y/dx^2 = (d/dx)\{[2x(x - 1) - x^2(1)]/(x - 1)^2\} = (d/dx)[(x^2 - 2x)/$

    $(x^2 - 2x + 1)] = [(2x - 2)(x^2 - 2x + 1) - (x^2 - 2x)(2x - 2)]/(x^2 - 2x + 1)$ =

    $2/(x - 1)^3$.

41. The velocity is $f'(t) = 3$ and $f'(1) = 3$ meters/second . The acceleration is $f''(t) = 0$, so $f''(1)$ is still 0 meters/second^2 .

45. The velocity is $f'(t) = -2 - 0.04t^3$ and $f'(0) = -2$ meters/second .
     The acceleration is $f''(t) = -0.12t^2$ and $f''(0) = 0$ meters/second^2 .

49. Marginal productivity, the derivative of the output function, is
     $20 - 2x$. When 5 workers are employed, marginal productivity is
     $20 - 2(5) = 10$. Thus, productivity would increase by 10 dollars
     per worker-hour.

53. This exercise is analogous to Example 12. The profit, $P(x)$, is
     $x(25 - 0.02x) - (4x + 0.02x^2)/(1 + 0.002x^3)$. Therefore, marginal
     profit is $P'(x) = 25 - 0.04x - [(4 + 0.04x)(1 + 0.002x^3) -$
     $(4x + 0.02x^2)(0.006x^2)]/(1 + 0.002x^3)^2 = [25 - 0.04x - (4 + 0.04x -$
     $0.016x^3 + 0.00004x^4)/(1 + 0.002x^3)^2]$ dollars/boot .

57. The rate of change of y with respect to x is the price of fuel in
     dollars per gallon or cents per liter. Other answers are possible.

61. The average rate of change is $\Delta y/\Delta x = [y(\Delta x) - y(0)]/\Delta x =$
     $[4(\Delta x)^2 - 2(\Delta x)]/\Delta x = 4\Delta x - 2$. By the quadratic function rule, the
     derivative at $x_0 = 0$ is $y'(0) = -2$, where $y'(x) = 8x - 2$. The
     average rate of change approaches the derivative as Δx gets smaller.
     For $\Delta x = 0.1$, 0.001 , and 0.000001 , the average rates of change
     are -1.6 , -1.996 , and -1.999996 , respectively.

65. The area, A , is ℓw , so $dA/dt = \ell(t)w'(t) + \ell'(t)w(t) =$
     $(3 + t^2 + t^3)(-1 + 4t) + (2t + 3t^2)(5 - t + 2t^2) = (10t^4 + 4t^3 + 12t^2 +$
     $22t - 3)$ cm^2/sec , which is the rate of change of area with respect to
     time.

69. Let $f(x) = ax^2 + bx + c$. Then $f'(x) = 2ax + b$ and the derivative of

this is $f''(x) = 2a$. Hence, $f''(x)$ is equal to zero when $a = 0$ — that

is, when $f(x)$ is a linear function $bx + c$.

73. (a) The linear equation is $V = V_0 + (\Delta V/\Delta t)t = 4000 + [(500 - 4000)/$

$10]t = 4000 - (350)t$.

(b) The slope is $\Delta V/\Delta t = -350$ (dollars/year) .

SECTION QUIZ

1. Tell what is wrong with this statement. Suppose $f(x) = -x^2 + 3x + 6$;

then $f'(x) = -2x + 3$. Evaluating, we get $f'(-1) = 5$, so $f''(-1) = 0$.

2. Compute $g''(4)$ for $g(x) = 4x - 3$ and for $g(y) = 2 - 3y^2$.

3. Compute the following derivatives:

(a) d^2u/dy^2 for $u = y^5 - y^3/2 + y$.

(b) $(d^2/dw^2)(w/3 + 3w^7/14 - 2w^4)$

(c) $g''(t)$ for $g(t) = 5t^5 + 4t^4 + 3t^3 + 2t^2 + t + 81$.

4. You and your spouse are planning to go on a werewolf hunt during the

next full moon. In preparation, you do a ballistics test and determine

the silver bullet's position as $6x^2 + 3x$ meters after x seconds.

Determine the acceleration of the bullet if it hits the werewolf 30

meters away.

5. Careless Christina, during the excitement of her twenty-first birthday,

mistakenly provided firecrackers for her birthday cake.

(a) Suppose firecrackers can expend 90x units of energy, where x

is the number of firecrackers. If the cake can absorb 15 units

of energy, write an equation relating the number of firecrackers

and the net energy liberated.

(b) Differentiate the function in (a) and give a physical interpretation

of the derivative.

ANSWERS TO PREREQUISITE QUIZ

1. (a) $10x + 1$

    (b) $3x^2 + 6x - 2$

    (c) $(-x^2 + 10x + 2)/(x^2 + 2)^2$

2. 7

3. The derivative of a function gives the slope of the tangent line.

ANSWERS TO SECTION QUIZ

1. One should find $f''(x)$ before evaluating; $f''(-1) = -2$.

2. 0 for $g(x)$; -6 for $g(y)$

3. (a) $20y^3 - 3y$

    (b) $9w^5 - 24w^2$

    (c) $100t^3 + 48t^2 + 18t + 4$

4. 27 meters/(second)^2

5. (a) $y = 90x - 15$ where $y =$ energy and $x =$ firecrackers

    (b) 90 ; the derivative is energy/firecracker

2.2   The Chain Rule

PREREQUISITES

1.   Recall how to differentiate polynomials, products, quotients, and

square roots (Section 1.4 and 1.5).

2.   Recall how to find limits (Section 1.2).

3.   Recall how to use functional notation (Section R.6).

PREREQUISITE QUIZ

1.   Differentiate the following functions with respect to x :

(a) $\sqrt{x}/(1 + x)$

(b) $(x^2 + x - 3)(x + 2)$

(c) $5x^4 - x^3/3$

2.   Compute the following limits:

(a) $\lim_{\Delta x \to 0} \{[(\Delta x)^2 + \Delta x]/\Delta x\}$

(b) $\lim_{x \to -1} [(x^2 + 4x + 3)/(x + 1)]$

3.   Let $f(x) = x^2 + 2x - 4$.

(a) Find $f(2)$.

(b) Find $f(y)$.

GOALS

1.   Be able to state and apply the chain rule.

2.   Be able to use the chain rule for solving word problems.

STUDY HINTS

1.   Power of a function rule . Don't bother memorizing this formula since

it will soon be covered by the chain rule and the rule $(d/dx)x^n = nx^{n-1}$.

Do learn how to apply it, though, as this is important preparation for

the chain rule.

2. <u>Composite function notation.</u> Become familiar with $(f \circ g)(x)$. This is the same as $f(g(x))$.

3. <u>Derivation of the chain rule.</u> You will probably not be expected to know the proof of the chain rule. It is much more important to understand how to apply the result.

4. <u>Chain rule.</u> Memorize $(f \circ g)'(x) = f'(g(x)) \cdot g'(x)$. Practice using this formula until you feel comfortable with it. It is probably the most important differentiation formula that you will learn. DO NOT forget the last factor.

5. <u>Leibniz notation.</u> The chain rule demonstrates the usefulness of the Leibniz notation. Notice how the du's appear to cancel in $dy/dx =$ $(dy/du) \cdot (du/dx)$. Remember that dy/dx is a derivative, not a fraction, but here they do behave like fractions.

6. <u>Shifting rule.</u> Geometrically, the shifting rule says that a horizontal displacement of a graph does not alter its slope. Don't memorize the formula. It is just a special case of the chain rule.

7. <u>Word problems.</u> Study Example 11 carefully. It is always a good idea to make a drawing, if possible. Many word problems will involve similar triangles. Notice how each of the rates are determined. Note also that the 8 feet did not enter into the solution of Example 11.

8. <u>Practical application.</u> The chain rule may be related to converting units For example, suppose we want to convert yards/second into meters/second. Let y be length in yards, let x be length in meters, and let t be time in seconds. Then $dy/dt = (dy/dx) \cdot (dx/dt)$. Here, dy/dx is the number of yards per meter.

SOLUTIONS TO EVERY OTHER ODD EXERCISE

1. Apply the power of a function rule to get $(d/dx)(x + 3)^4 = 4(x + 3)^3$ x $(d/dx)(x + 3) = 4(x + 3)^3$.

5. Apply the power of a function rule and the product rule to get $(d/dx)[(x^2 + 8x)^3 x] = [(d/dx)(x^2 + 8x)^3]x + (x^2 + 8x)^3(1) = 3x(x^2 + 8x)^2$ x $(d/dx)(x^2 + 8x) + (x^2 + 8x)^3 = 3x(x^2 + 8x)^2(2x + 8) + (x^2 + 8x)^3 = (x^2 + 8x)^2(7x^2 + 32x)$.

9. Apply the power of a function rule and the product rule to get $(d/dy)[(y + 1)^3(y + 2)^2(y + 3)] = [(d/dy)(y + 1)^3] \cdot (y + 2)^2(y + 3) + (y + 1)^3 \cdot (d/dy)[(y + 2)^2(y + 3)] = 3(y + 1)^2(y + 2)^2(y + 3) \cdot (d/dy)(y + 1) + (y + 1)^3 \cdot \{[(d/dy)(y + 2)^2] \cdot (y + 3) + (y + 2)^2 \cdot (d/dy)(y + 3)\} = 3(y + 1)^2(y + 2)^2(y + 3) + 2(y + 1)^3(y + 2)(y + 3) + (y + 1)^3(y + 2)^2 = (y + 1)^2(y + 2)(6y^2 + 26y + 26)$.

13. By definition, $(f \circ g)(x) = f(g(x)) = f(x^3) = (x^3 - 2)^3$. Also, $(g \circ f)(x) = g(f(x)) = g((x - 2)^3) = ((x - 2)^3)^3 = (x - 2)^9$.

17. There is no unique answer. One solution is to let $h(x) = f(g(x))$ with $f(x) = \sqrt{x}$ and $g(x) = 4x^3 + 5x + 3$. Notice that the choice of variable here differs slightly from the answer in the text.

21. We can compute $h(x) = f(g(x)) = (x^2 - 1)^2 = x^4 - 2x^2 + 1$ and then differentiate directly to get $h'(x) = 4x^3 - 4x = 4x(x^2 - 1)$. On the other hand, the chain rule gives $f'(g(x)) \cdot g'(x) = 2u \cdot 2x = 2(x^2 - 1)(2x) = 4x(x^2 - 1)$.

25. Let $f(u) = u^3$ and $u = x^2 - 6x + 1$. Then $(d/dx)f(g(x)) = f'(g(x)) \cdot g'(x)$, where $u = g(x)$. Thus, the derivative is $3u^2(2x - 6) = 6(x^2 - 6x + 1)^2(x - 3)$.

29. Let $f(x)$ denote the function. Recall that if $u = g(x)$, then
     $(d/dx)f(g(x)) = f'(g(x)) \cdot g'(x)$. Now apply the chain rule twice.
     First, let $f(u) = u^2$ and $u = (x^2 + 2)^2 + 1$, so $f'(x) =$
     $2u[(d/dx)((x^2 + 2)^2 + 1)]$. Now, let $f(u) = u^2 + 1$ and $u = x^2 + 2$,
     so $(d/dx)[(x^2 + 2)^2 + 1] = 2u(2x) = 2(x^2 + 2)(2x)$. Thus, $f'(x) =$
     $2[(x^2 + 2)^2 + 1]2(x^2 + 2)(2x) = 8x(x^2 + 2)[(x^2 + 2)^2 + 1]$.

33. If $f(x)$ is the given function and $u = g(x)$, then $(d/dx)f(g(x)) =$
     $f'(g(x)) \cdot g'(x)$. So let $f(u) = \sqrt{u}$ and $u = 4x^5 + 5x^2$ to get
     $f'(x) = (1/2\sqrt{u})(20x^4 + 10x) = 5x(2x^3 + 1)/\sqrt{4x^5 + 5x^2}$.

37. (a) It would seem reasonable that $(f \circ g \circ h)(x)$ can be defined as
          $[f \circ (g \circ h)](x)$. This is $f((g \circ h)(x))$, but since $(g \circ h)(x) =$
          $g(h(x))$, it becomes $(f \circ g \circ h)(x) = f(g(h(x)))$.

     (b) Let $u = g(h(x))$, then the derivative of $f \circ g \circ h$ is $f'(u) \cdot u'(x)$.
          Applying the chain rule to u gives $u'(x) = g'(h(x)) \cdot h'(x)$.
          Therefore, $(f \circ g \circ h)'(x) = f'(g(h(x))) \cdot g'(h(x)) \cdot h'(x)$.

41. The Leibniz notation is very useful here. We write $dK/dt = (dK/dv) \times$
     $(dv/dt) = mv \cdot (dv/dt)$. Since dv/dt is the acceleration, we can sub-
     stitute the appropriate values, yielding $dK/dt = (10)(30)(5)$
     $(gram - cm^2/sec^2)/sec = 1500$ gram $- cm^2/sec^3$.

45. The velocity is the derivative of the position function. $(d/dt)((t^2 + 4)^5)$
     $5(t^2 + 4)^4(2t)$. At $t = -1$, the velocity is -6250 .

49. If $f(x) = (x^4 + 10x^2 + 1)^{98}$, the power of a function rule gives
     $f'(x) = 98(x^4 + 10x^2 + 1)^{97}(4x^3 + 20x) = 392(x^3 + 5x)(x^4 + 10x^2 + 1)^{97}$.
     Next, the product rule along with the power of a function rule gives
     $f''(x) = 392(x^3 + 5x) \cdot (d/dx)(x^4 + 10x^2 + 1)^{97} + 392(x^4 + 10x^2 + 1)^{97} \times$
     $(d/dx)(x^3 + 5x) = 392(x^4 + 10x^2 + 1)^{96}[391x^6 + 3915x^4 + 53x^2 + 9700x + 5]$

53. Applying the chain rule once gives $(d^2/dx^2)(u^n) = (d/dx)(nu^{n-1}(du/dx))$.
    Now use the constant multiple and product rules to get $n(d/dx)(u^{n-1}(du/dx))$,
    which becomes $n[(n-1)u^{n-2}(du/dx)(du/dx) + u^{n-1}(d^2u/dx^2)]$. Therefore,
    $(d^2/dx^2)(u^n) = nu^{n-2}[u(d^2u/dx^2) + (n-1)(du/dx)^2]$.

57. By the chain rule, $(f \circ g)'(x) = f'(g(x)) \cdot g'(x)$. Then the product rule
    is applied: $(f \circ g)''(x) = [f''(g(x)) \cdot g'(x)] g'(x) + f'(g(x)) \cdot g''(x) =$
    $f''(g(x)) [g'(x)]^2 + f'(g(x)) g''(x)$.

SECTION QUIZ

1. Given that $f(2) = 2$, $g(2) = 4$, $f(4) = 3$, $g(4) = 5$, $f'(2) = -1$,
    $g'(2) = -2$, $f'(4) = 4$, $g'(4) = -4$, what is $(d/dx)f(g(x))$ at $x = 2$?

2. $(3x + 1)^3 = 27x^3 + 27x^2 + 9x + 1$, so the derivative is $81x^2 + 54x + 9$.
    On the other hand, the power of a function rule tells us that the deriva-
    tive is $3(3x + 1)^2 = 3(9x^2 + 6x + 1) = 27x^2 + 18x + 3$. Why aren't the
    derivatives equal?

3. If $f(t) = (t + 1)^2$ and $g(t) = t^2 + 1$, what is $(g \circ f)(x)$? $(f \circ g)(x)$?

4. Find dy/dx for the following functions:

    (a) $y = \sqrt{(2x - 3)^3 + 1}$

    (b) $y = [(x^3 + 1)^4 - 5]/[2 - (3x^2 - 2)^3]$

5. Find a formula for the second derivative of $f(g(x))$ with respect to x .

6. A jack-in-the-box suddenly springs up at 10 cm/sec . It is located one
    meter from a lamp on the floor. The toy casts a scary shadow on the wall
    3 meters from the lamp. How fast is the shadow enlarging when the
    jack-in-the-box has risen 5 cm . ?

7. Wrong-way Willie sometimes gets absent-minded while driving his new
    Porsche. He has a tendency to drive on the opposite side of the street
    and to drive in the wrong direction on one-way streets. It is estimated
    that Willie drives illegally for $0.1x$ mile in each x total miles
    driven. On the average, he sideswipes w^2 cars after w miles of
    illegal driving. It is also known that he always drives at 30 miles/hour

    (a) Use the chain rule to compute the rate of sideswiping per mile of
         total driving.

    (b) Compute the rate of sideswiping per hour of total illegal driving.

ANSWERS TO PREREQUISITE QUIZ

1.   (a) $(1 - x)/2\sqrt{x}(1 + x)^2$

     (b) $3x^2 + 6x - 1$

     (c) $20x^3 - x^2$

2.   (a) 1

     (b) 2

3.   (a) 4

     (b) $y^2 + 2y - 4$

ANSWERS TO SECTION QUIZ

1.   -8

2.   In the second method, we forgot to differentiate the function within
     the parentheses.

3. $(g \circ f)(x) = (x + 1)^4 + 1$; $(f \circ g)(x) = (x^2 + 2)^2$.

4.   (a) $3(2x - 3)^2/\sqrt{(2x - 3)^3 + 1}$

     (b) $\{12x^2(x^3 + 1)^3[2 - (3x^2 - 2)^3] + 18x(3x^2 - 2)^2[(x^3 + 1)^4 - 5]\}/$
          $[2 - (3x^2 - 2)^3]^2$

5. $f''(g(x)) \cdot (g'(x))^2 + f'(g(x)) \cdot g''(x)$

6.   30 cm/sec

7.   (a)   w/10

     (b)   3w

2.3   Fractional Powers and Implicit Differentiation

PREREQUISITES

1.   Recall how to differentiate rational functions, especially those which
     require the use of the chain rule (Section 2.2).

PREREQUISITE QUIZ

1.   Differentiate $\sqrt{2x}$.

2.   Differentiate $(x^2 + 5)^6$.

3.   Suppose $g(t) = t^2 + 3$ and $f(x) = 2x$;   what is $(d/dx)g(f(x))$?

GOALS

1.   Be able to differentiate functions with fractional exponents.

2.   Be able to use the method of implicit differentiation.

STUDY HINTS

1.   Rational power rule. This is just an extension of the power rule for
     integers. The power rule is now valid for all rational numbers for
     which the derivative is defined.

2.   Rational power of a function rule. As with the rational power rule,
     this is just an extension of a previously learned rule - the power of
     a function rule for integers.

3.   Implicit differentiation. One of the common mistakes in applying this
     method is forgetting that y is a function of x . Thus, $(d/dx)(y^2) =$
     $2yy'$, not just $2y$; the latter would be $(d/dy)y^2$. Be sure you
     understand the method of implicit differentiation. You will probably
     get one of these problems on your exams.

SOLUTIONS TO EVERY OTHER ODD EXERCISE

1. The rational power rule gives $(d/dx)(10x^{1/8}) = (5/4)x^{-7/8} = 5/4x^{7/8}$.

5. By the rational power rule, $(d/dx)(3x^{2/3} - (5x)^{1/2}) = 2x^{-1/3} -$
    $(1/2)(5x)^{-1/2} \cdot 5 = 2/x^{1/3} - 5/2\sqrt{5x}$. Actually, the rational power of
    a function rule was used to differentiate the second term.

9. By the rational power of a function rule, $(d/dx)[(x^5 + 1)^{7/9}] =$
    $(7/9)(x^5 + 1)^{-2/9}(5x^4) = 35x^4/9(x^5 + 1)^{2/9}$.

13. Using the quotient rule with the rational power of a function rule, we
    get $(d/dx)\left[\sqrt{x^2 + 1}/\sqrt{x^2 - 1}\right] = (d/dx)[(x^2 + 1)/(x^2 - 1)]^{1/2} =$
    $(1/2)[(x^2 + 1)/(x^2 - 1)]^{-1/2} \cdot [(2x)(x^2 - 1) - (x^2 + 1)(2x)]/(x^2 - 1)^2 =$
    $-2x/(x^2 + 1)^{1/2}(x^2 - 1)^{3/2}$.

17. By the quotient rule with the rational power rule, we get $(d/dx)[\sqrt{x}/$
    $(3 + x + x^3)] = [(1/2)x^{-1/2}(3 + x + x^3) - \sqrt{x}(1 + 3x^2)]/(3 + x + x^3)^2 =$
    $[(3 + x + x^3 - 2x - 6x^3)/2\sqrt{x}]/(3 + x + x^3)^2 = (3 - x - 5x^3)/2\sqrt{x}(3 + x + x^3)^2$.

21. Combining the quotient rule with the rational power rule, we get
    $(d/dx)[\sqrt[3]{x}/(x^2 + 2)] = [(1/3)x^{-2/3}(x^2 + 2) - \sqrt[3]{x}(2x)]/(x^2 + 2)^2 =$
    $[(x^2 + 2 - 6x^2)/3x^{2/3}]/(x^2 + 2)^2 = (2 - 5x^2)/3x^{2/3}(x^2 + 2)^2$.

25. By the rational power of a function rule, $(d/dx)(x^2 + 5)^{7/8} =$
    $(7/8)(x^2 + 5)^{-1/8}(2x) = 7x/4(x^2 + 5)^{1/8}$.

29. The sum rule and the rational power rule give $f'(x) = (3/11)x^{-8/11} -$
    $(1/5)x^{-4/5}$.

33. Applying the rational power of a function rule gives $\ell'(x) =$
    $(1/2)[(x^{1/2} + 1)/(x^{1/2} - 1)]^{-1/2}[(1/2)x^{-1/2}(x^{1/2} - 1) -$
    $(x^{1/2} + 1)(1/2)x^{-1/2}](x^{1/2} - 1)^2 = -1/2\sqrt{x}(\sqrt{x} - 1)^{3/2}(\sqrt{x} + 1)^{1/2}$.

37. (a) Differentiating with respect to x , we get $4x^3 + 2y(dy/dx) +$
    $dy/dx = 0$, i.e., $(dy/dx)(2y + 1) = - 4x^3$. Hence $dy/dx =$
    $-4x^3/(2y + 1)$.

    (b) $(dy/dx)\big|_{x=1,y=1} = (-4)/(2 + 1) = -4/3$.

    (c) $y^2 + y + (x^4 - 3) = 0$ implies $y = [-1 \pm \sqrt{1 - 4(1)(x^4 - 3)}]/2 =$
    $[-1 \pm \sqrt{13 - 4x^4}]/2$, so $dy/dx = \pm[1(-16x^3)/4\sqrt{13 - 4x^4}] =$
    $\pm[4x^3/\sqrt{13 - 4x^4}]$. Note that $2y + 1 = -1 \pm \sqrt{13 - 4x^4} + 1 =$
    $\pm\sqrt{13 - 4x^4}$. Hence $dy/dx = -4x^3/(2y + 1)$. The answer checks.

41. Using implicit differentation, $4x^3 + 4y^3(dy/dx) = 0$. When $x = y = 1$,
    we get $4 + 4(dy/dx) = 0$ or $dy/dx = -1$. The equation of the tangent
    line is $y = 1 + (-1)(x - 1) = -x + 2$.

45. $dy/dx = -2x/2\sqrt{1 - x^2} = -x/\sqrt{1 - x^2}$, so $(dy/dx)\big|_{x = \sqrt{3}/2} = -(\sqrt{3}/2)/(1/2) =$
    $-\sqrt{3}$. Thus, the equation of the tangent line is $y = 1/2 - \sqrt{3}(x - \sqrt{3}/2)$
    or $y = -\sqrt{3}x + 2$.

49. Use the linear approximation $f(x_0 + \Delta x) \approx f(x_0) + f'(x_0)\Delta x$. Let $f(x) =$
    $x^{1/4}$, so $f'(x) = (1/4)x^{-3/4}$, and so $f(15.97) = f(16 - 0.03) \approx$
    $f(16) + f'(16)(-0.03) = 2 + (1/4)(1/8)(-0.03) = 2 - 0.03/32 = 1.9990625$.

53. Applying the rational power rule with the chain rule yields $dM/dx =$
    $24(4)(2 + x^{1/3})^3(1/3)x^{-2/3} = [32(2 + x^{1/3})^3/x^{2/3}]$ kg/unit distance.

57. The rate of change of period with respect to tension is $dP/dT =$
    $(1/2)(32/T)^{-1/2}(-32/T^2) = -2\sqrt{2}/T^{3/2}$; at $T = 9$, $dP/dT = -2\sqrt{2}/27$
    seconds/pound .

SECTION QUIZ

1. $x^2y + 2y = (x + y)^3 - 5$ describes a differentiable curve. Find the tangent line to the curve at $(1,1)$.

2. Find dy/dx if $y = (x^{1/2} + x^{1/3})/(1 + \sqrt{x})^{2/3}$.

3. Find d^2f/dw^2 if $f = (3w^{1/3} - \sqrt{w^3})/w^{2/5}$.

4. For the curve in Question 1, compute d^2y/dx^2 at $(1,1)$.

5. Four eyes Frankie could hardly see even with his glasses on. Thus, when he got into his motor boat, he steered it along the path described by $(x^2y + 4)^{3/2} = 22x/\sqrt{y + 2} + 2x^3 + 11x^{1/2}y^2/2$. He kept this up for an hour until he lost control at $(4,2)$ and went sailing off along the tangent line. What is this tangent line?

ANSWERS TO PREREQUISITE QUIZ

1. $1/\sqrt{2x}$

2. $12x(x^2 + 5)^5$

3. $8x$

ANSWERS TO SECTION QUIZ

1. $9y + 10x = 19$

2. $[(3/2 + \sqrt{x}/\sqrt[3]{x^2})(1 + \sqrt{x}) - (\sqrt{x} + \sqrt[3]{x})]/3\sqrt{x}(1 + \sqrt{x})^{5/3}$.

3. $48/225w^{31/15} - 11/100w^{9/10}$

4. $-70/81$

5. $211y + 207x = 1250$

2.4 Related Rates and Parametric Curves

PREREQUISITES

1.   Recall how to differentiate by using the chain rule (Section 2.2).

PREREQUISITE QUIZ

1.   Differentiate $\sqrt{x^2 - 3}$.

2.   Differentiate $(x - 3)^5(x^2 + 1)^4$.

GOALS

1.   Be able to solve related rates word problems.

2.   Be able to calculate the slope of a parametric curve.

3.   Be able to sketch simple parametric curves.

STUDY HINTS

1.   <u>Related rates</u>. Don't forget to use the chain rule when you differentiate
     x and y with respect to t . You may want to substitute $x = f(t)$
     and $y = g(t)$ to help remind yourself to use the chain rule.

2.   <u>Parametric curves</u>. Two curves may appear to be related by the same
     equation and yet they are not the same curve. For example, $x = t^6$ and
     $y = t^3$ are related by $y^2 = x$, which is an entire parabola; y may
     take on negative values. Compare this with Example 2.

3.   <u>Finding an xy-relationship</u>. Sometimes it is useful to find an equation
     involving only x and y when you are asked to sketch a parametric
     curve. This can often be done by solving for t in one equation and
     substituting into the other.

4.   <u>Slopes of parametric curves</u>. Remember that $dy/dx = (dy/dt)/(dx/dt)$.
      As with the chain rule, the dt's appear to cancel, but remember they
      are not really fractions.

5.   <u>Word problems</u>. Many word problems involve related rates. Draw a pic-
      ture, if possible. Look for a relationship between the variables.
      Sometimes you will have to derive a relationship as in Example 6.
      Differentiate both sides of your relationship with respect to time.
      Finally, substitute in the given values. A few minutes spent in
      studying Examples 6, 7, and 8 should prove worthwhile.

SOLUTIONS TO EVERY OTHER ODD EXERCISE

1.   Differentiate x and y with respect to t as in Example 1 to get
      $2x(dx/dt) - 2y(dy/dt) = 0$, so $2y(dy/dt) = 2x(dx/dt)$ or $dy/dt =$
      $(x/y)(dx/dt)$.

5.   Differentiation of x and y with respect to t yields $dx/dt +$
      $2y(dy/dt) = dy/dt$, so $(1 - 2y)(dy/dt) = dx/dt$ or $dy/dt =$
      $(dx/dt)/(1 - 2y)$.

9.   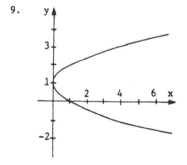 $y = 1 - t$ implies $t = 1 - y$, so the
      curve is $x = (1 - y)^2$, which is a
      parabola symmetric about $y = 1$.

13.   The slope of the tangent line is $(dy/dt)/(dx/dt) = 3t^2/2t = 3t/2$. Thus,
      at $t = 5$, the slope is $15/2$, $x = 25$, and $y = 125$. Hence, the
      equation of the line is $y = 125 + (15/2)(x - 25)$ or $y = (15x - 125)/2$.

17. Differentiating with respect to t , we get $(dx/dt)y + (dy/dt)x = 0$.
Substituting $x = 8$ and $y = 1/2$ yields $(1/2)(dx/dt) + 8(dy/dt) = 0$,
and so $dy/dt = (1/2)(dx/dt)/(-8) = -(dx/dt)/16$.

21. Let $r(t)$ be the radius at time t and let $h(t)$ be the height at
time t . We have volume $= V = 1000 = \pi(r(t))^2 h(t)$. Differentiating
with respect to t , we get $0 = \pi[2r(t)r'(t)h(t) + (r(t))^2 h'(t)]$.
At the instance when $r(t) = 4$, $r'(t) = 1/2$ and $h(t) = 1000/\pi(4)^2 =$
$62.5/\pi$. Substituting in all of these known values, we get
$0 = \pi[2(4)(1/2)(62.5/\pi) + (4)^2 h'(t)]$, i.e., $h'(t) = -250/16\pi =$
$-(125/8\pi)$ cm/sec .

25. (a) By the distance formula, we have $\sqrt{x^2 + y^2} = 2\sqrt{x^2 + (y - 1)^2}$.
Squaring and rearranging yields $3x^2 + 3(y - 4/3)^2 = 4/3$, which
is a circle centered at $(0, 4/3)$ with radius $2/3$.

(b) We want to know dy/dt at $(0, 2/3)$. By implicit differentiation,
we have $6x(dx/dt) + 6(y - 4/3)(dy/dt) = 0$. At $(0, 2/3)$, the
equation is $4(dy/dt) = 0$, so $dy/dt = 0$.

(c) Rearranging the equation in part (b), we get $(dy/dt)/(dx/dt) = 1 =$
$-x/(y - 4/3)$; therefore, we need $x = 4/3 - y$. Substitute into
the equation in part (a): $3(4/3 - y)^2 + 3(y - 4/3)^2 = 6(y - 4/3)^2 =$
$4/3$, so $y - 4/3 = \pm 2/3\sqrt{2}$. Therefore, $y = (\pm\sqrt{2} + 4)/3$, and
the points are $(\sqrt{2}/3, (-\sqrt{2} + 4)/3)$ and $(-\sqrt{2}/3, (\sqrt{2} + 4)/3)$.

29. Let ℓ and w denote the length of the rectangle's sides. We want
to know what $d\ell/dw$ or $dw/d\ell$ is when $\ell = w = 5$. Differentiating
$\ell w = 25$ with respect to ℓ yields $w + \ell(dw/d\ell) = 0$ or $dw/d\ell =$
$-w/\ell = -5/5 = -1$. On the other hand, differentiating with respect to
w , $(d\ell/dw)w + \ell = 0$ yields $-\ell/w = d\ell/dw = -5/5 = -1$.

33. Let y be the rainfall rate, R be the radius of the tank, and H be
the height of the tank. The other variables have the same meaning as
in Example 8. dV/dt becomes $\pi R^2 y$, r/h = R/H , and r = Rh/H .
Therefore, $V = \pi r^2 h/3 = \pi(Rh/H)^2 h/3 = \pi R^2 h^3/3H^2$. Differentiation
yields $dV/dt = \pi R^2 h^2(dh/dt)/H^2 = \pi R^2 y$; therefore, $(dh/dt)/y =$
H^2/h^2 . But from r/h = R/H , we get $R^2/r^2 = H^2/h^2$ or $H^2/h^2 =$
$\pi R^2/\pi r^2 = (dh/dt)/y$, which is the desired result.

SECTION QUIZ

1. A curve is described by $x = t^4$, $y = t^2$ and another curve is described
by $x = t^6$, $y = t^3$. Sketch the two curves.

2. (a) If $x = t^3 - 2t^2$ and $y = t^2 - 4$, what is dy/dx whenever the
curve crosses the x-axis?

(b) At what points is the tangent line horizontal?

(c) At what points is the tangent line vertical?

3. (a) Airbelly Alice just got a job perfectly suited for her rotund
tummy. Her new job is blowing up balloons for the circus. If
the spherical balloon inflates at a rate of 5 cc/min., how fast
is the diameter increasing when the radius is $\sqrt{5}$ cm. ?

(b) Airbelly Alice's tummy has the shape of a circular cylinder. If
all of the air used for blowing up the balloons comes from her
belly which is 10 cm. high, how fast is Alice's waistline decreasing
when her tummy is 15 cm. in radius?

4.   Queer Mr. Q, who enjoys giraffeback riding, needed a new fence to pre-
      vent his giraffe from running away. He ordered the fence installer to
      put up the fence according to the following specifications:

      For $-1 \leqslant t \leqslant 3$, $x = t - 1$ and $y = 5 - t^2 - 2t$. Then for

      $3 \leqslant t \leqslant 7$, $x = 5 - t$ and $y = (5 - t)^4 - 14$.

      (a) Suppose at $t = 4$, the giraffe runs away along the tangent line.
           What path does it follow?

      (b) The frightened fence installer ran off at a perpendicular at $t = 4$.
           He is sprinting at 9 kilometers/hour , while the giraffe is
           running at 12 kilometers/hour . How fast is their distance
           increasing after 15 minutes?

      (c) Make a sketch of the completed giraffe pen.

ANSWERS TO PREREQUISITE QUIZ

1. $x\sqrt{x^2 - 3}$

2. $(x - 3)^4 (x^2 + 1)^3 (13x^2 - 24x + 5)$

ANSWERS TO SECTION QUIZ

1.

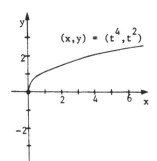
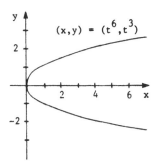

2. (a)   1 when t = 2 and −1/5 when t = −2 .

    (b)   At no points

    (c)   (−32/27,−20/9)   when   t = 4/3 .

3. (a)   1/2π cm/min

    (b)   The radius decreases at   1/60π cm/min., so her waistline is
          decreasing at 1/30 cm/min .

4. (a)   y − 4x + 17 = 0

    (b)   15 km/hr

    (c)

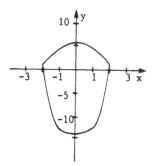

2.5   Antiderivatives

PREREQUISITES

1.   Recall how to differentiate a polynomial (Section 1.4).

2.   Recall how to differentiate a composite function (Section 2.2).

3.   Recall how position and velocity are related by the derivative (Section 1.1)

PREREQUISITE QUIZ

1.   Differentiate $x^{48} - 5x^5 + x^3 - 3x + 25$.

2.   What is $(d/dx)f(g(x))$?

3.   Differentiate $(3x + 2)^4$.

4.   Suppose $y = 3x^3 - 2x^2 + 4x - 4$ describes a particle's position y at

time x .

(a)   What is dy/dx ?

(b)   What is the physical interpretation of dy/dx ?

GOALS

1.   Be able to find antiderivatives for polynomials and simple composite

functions.

2.   Be able to interpret the meaning of an antiderivative.

STUDY HINTS

1.   Antiderivatives. Remember that an antiderivative is not unique unless an

extra condition is given. Always remember to include the arbitrary con-

stant. It is a common mistake to forget the arbitrary constant.

2.   Power rule. $n = -1$ is excluded because the antiderivative would require

division by 0 .

3. Polynomial rule. This rule incorporates the sum rule, the constant
    multiple rule, and the power rule. You should learn the basic parts
    well and be able to derive the polynomial rule by yourself.

4. Antidifferentiating composite functions. A systematic method will be
    introduced in Chapter 7. For now, think of the quantity inside the
    parenthesis as a single variable when you guess an antiderivative.
    Then differentiate as in Example 7.

5. Physical interpretation. To help you understand Example 11, recall
    that differentiation yields a rate. Antidifferentiation will yield
    the original function. Therefore, antidifferentiating the water flow
    rate should give the total amount of water.

SOLUTIONS TO EVERY OTHER ODD EXERCISE

1. Apply the polynomial rule for antidifferentiation to get $F(x) =$
    $x^2/2 + 2x + C$.

5. Apply the power rule for antidifferentiation to get $F(t) =$
    $t^{-3+1}/(-3 + 1) + C = -1/2t^2 + C$.

9. By the result of Example 10, the position function is $F(t) = \int v dt$.
    By the polynomial rule for antidifferentation, $F(t) = 4t^2 + 2t + C$.
    Therefore, $F(0) = 0$ implies $C = 0$, and so $F(1) = 4(1)^2 + 2(1) +$
    $0 = 6$.

13. Using the polynomial rule for antidifferentiation, $F(x) = (3/2)x^2 + C$.

17. Use the formula $\int(ax + b)^n dx = (ax + b)^{n+1}/a(n + 1) + C$. Here, $a = 1$
    and $b = 1$, so the general antiderivative is $F(x) = 2(x + 1)^{3/2}/3 + C$.

21. The acceleration is 9.8 near the earth's surface, so $v = 9.8t + C$,
     which is v_0 at $t = 0$. Thus, $v = 9.8t + v_0$, and the position
     function becomes $x = 4.9t^2 + v_0 t + D$. $x = x_0$ at $t = 0$, so $x =$
     $4.9t^2 + v_0 t$ x_0 . Since $v_0 = 1$, we have $v = 9.8t + 1$, and since
     $x_0 = 2$, $x = 4.9t^2 + t + 2$.

25. It is not true. For a counterexample, take $f(x) = x$ and $g(x) = 1$.
     Then $\int f(x)g(x)dx = \int x dx = x^2/2$. (For simplicity, let all constants be
     $C = 0$.) Now $\int f(x) dx = x^2/2$ and $\int g(x)dx = x$, so $[\int f(x)dx]g(x) +$
     $f(x)[\int g(x)dx] = (x^2/2)(1) + x(x) = (3/2)x^2$. This is not equal to
     $\int f(x)g(x)dx = x^2/2$.

29. By the polynomial rule for antidifferentiation, $\int (x^2 + 3x + 2)dx =$
     $x^3/3 + 3x^2/2 + 2x + C$.

33. Using $\int (ax + b)^n dx = (ax + b)^{n+1}/a(n + 1) + C$, $F(t) = (8t + 1)^{-1}/$
     $8(-1) + C = -1/8(8t + 1) + C$.

37. Using the polynomial rule for antidifferentiation, $\int (1/x^4 + x^4)dx =$
     $\int (x^{-4} + x^4)dx = x^{-3}/(-3) + x^5/5 + C = -1/3x^3 + x^5/5 + C$.

41. By using the polynomial rule for antidifferentiation, $\int (x^3 + 3x)dx =$
     $x^4/4 + 3x^2/2 + C$.

45. Use the formula $\int (ax + b)^n dx = (ax + b)^{n+1}/a(n + 1) + C$ to get
     $\int (8x + 3)^{1/2}dx = (8x + 3)^{3/2}/8(3/2) + C = (8x + 3)^{3/2}/12 + C$.

49. Simplification gives $\int [(\sqrt{x - 1} + 3)/(x - 1)^{1/2}]dx = \int [1 + 3(x - 1)^{-1/2}]dx$.
     Using the sum rule for antidifferentiation and $\int (ax + b)^n dx =$
     $(ax + b)^{n+1}/a(n + 1) + C$, the antiderivative is $x + 3(x - 1)^{1/2}/(1/2) +$
     $C = x + 6\sqrt{x - 1} + C$.

53. From Example 4, $x = 4.9t^2 + v_0 t + x_0$ where $v_0 = 10$ meters/sec and
    $x_0 = 0$,   i.e., $x = 4.9t^2 + 10t$.   We want to find t such that
    $x = 150 = 4.9t^2 + 10t$.   Using the quadratic formula in solving $4.9t^2 +$
    $10t - 150 = 0$,   we find $t = \left[-10 + \sqrt{10^2 + 4(150)(4.9)}\right]/2(4.9) = 4.6$ sec .

57. From Example 4, we have the formula $x = 4.9t^2 + v_0 t + x_0$,   where v_0
    is the <u>downward</u> velocity which is -19.6 for this problem. We want
    the time when $x = x_0$,   so we solve $0 = 4.9t^2 - 19.6t = t(4.9t - 19.6)$.
    This has solutions 0 and 4 ,   but 0 does not make sense, so $t = 4$
    seconds.

61. (a) By the power of a function rule, $(d/dx)(x^4 + 1)^{20} =$
        $20(x^4 + 1)^{19}(4x^3) = 80(x^4 + 1)^{19}x^3$.

    (b) By the sum rule for antidifferentiation and part (a), the integral
        is $(x^4 + 1)^{20}/80 + 9x^{5/3}/5 + C$.

65. By the polynomial rule for antidifferentiation, $F(x) = x^4/4 + x^3 + 2x + C$.
    $F(0) = 1$ implies that $C = 1$,   so $F(x) = x^4/4 + x^3 + 2x + 1$.

SECTION QUIZ

1. Calculate the following antiderivatives:

    (a) $\int (x + 3)(x + 1)dx$

    (b) $\int [(x^3 - 3x^2)/x^{3/2}] dx$

    (c) $\int (-2t - 5)dt$

    (d) $\int -389 \, dy$

2. (a) Differentiate $(x^4 + 4x)^3$.

    (b) Find the antiderivative $F(z)$ of $f(z) = 2(z^4 + 4z)^2(z^3 + 1)$ such
        that $F(0) = 5$.

3.   Evaluate $\int (3t + 7)^5 dt$.

4.   A rich stranger has just dropped his gold plated credit card into a
      tank of lobsters. Fearful of being pinched, he hires you to reach in
      and retrieve his credit card. He offers you $1,000, but his daddy
      always told him, "Time is money." Thus, he will decrease your pay at
      a rate of 50t dollars per minute, i.e., after x minutes, you will
      lose $\int_0^x 50t \, dt$ dollars. How much time can you use to retrieve the
      card and still earn $800 ?

ANSWERS TO PREREQUISITE QUIZ

1. $48x^{47} - 25x^4 + 3x^2 - 3$

2. $f'(g(x)) \cdot g'(x)$

3. $12(3x + 2)^3$

4.   (a) $9x^2 - 4x + 4$

      (b) Velocity

ANSWERS TO SECTION QUIZ

1.   (a) $x^3/3 + 2x^2 + 3x + C$

      (b) $2x^{5/2}/5 - 2x^{3/2} + C$

      (c) $-t^2 - 5t + C$

      (d) $-389y + C$

2.   (a) $12(x^4 + 4x)^2(x^3 + 1)$

      (b) $F(z) = (z^4 + 4z)^3/6 + 5$

3. $(3t + 7)^6/18 + C$

4. $\sqrt{8}$ minutes

2.R Review Exercises for Chapter 2

SOLUTIONS TO EVERY OTHER ODD EXERCISE

1. Apply the power of a function rule to get $3(6x + 1)^2 \cdot 6 = 18(6x + 1)^2$.

5. Using the power rule, $(d/dx)(6/x) = (d/dx)(6x^{-1}) = -6/x^2$.

9. Combine the quotient and chain rules to get $[13(x^2 + 1)^{12}(2x)(x^2 - 1)^{14} -$
    $(x^2 + 1)^{13}(14)(x^2 - 1)^{13}(2x)] / (x^2 - 1)^{28} = (-2x^3 - 54x)(x^2 + 1)^{12} /$
    $(x^2 - 1)^{15}$.

13. By using the quotient rule, the derivative is $[A'(x)D(x) - A(x)D'(x)] /$
    $[D(x)]^2 = [(3x^2 - 2x - 2)(x^2 + 8x + 16) - (x^3 - x^2 - 2x)(2x + 8)] /$
    $(x^2 + 8x + 16)^2 = [(3x^2 - 2x - 2)(x + 4) - (x^3 - x^2 - 2x)(2)] / (x + 4)^3 =$
    $(x^3 + 12x^2 - 6x - 8) / (x + 4)^3$.

17. Recall that the tangent line is $y = f(x_0) + f'(x_0)(x - x_0)$. By the
    power of a function rule, $f'(x) = (1/3)[A(x)]^{-2/3}A'(x) = (x^3 - x^2 - 2x)^{-2/3} \times$
    $(3x^2 - 2x - 2)/3$, so $f'(1) = -(1/3)(1/4)^{1/3} = -\sqrt[3]{2}/6$. Also, $f(1) = -1$,
    so the tangent line is $y = -1 - \sqrt[3]{2}(x - 1)/6$.

21. By the rational power rule, $f'(x) = (5/3)x^{2/3}$.

25. Applying the quotient and rational power rules gives $f'(x) = [(3/2)x^{1/2}$
    $(1 - x^{3/2}) - (-3/2)x^{1/2}(1 + x^{3/2})] / (1 - x^{3/2})^2 = 3\sqrt{x}/(1 - x^{3/2})^2$.

29. Use the quotient rule to get $f'(x) = [(1)(x^2 + 2bx + c) - (x - a)(2x + 2b)] /$
    $(x^2 + 2bx + c)^2 = (-x^2 + 2ax + c + 2ab)/(x^2 + 2bx + c)^2$;
    $f''(x) = [(-2x + 2a)(x^2 + 2bx + c)^2 - (-x^2 + 2ax + c + 2ab)(2) \times$
    $(x^2 + 2bx + c)(2x + 2b)] / (x^2 + 2bx + c)^4 = [(-2x + 2a)(x^2 + 2bx + c) -$
    $(-x^2 + 2ax + c + 2ab) \cdot (4x + 4b)] / (x^2 + 2bx + c)^3 = [2x^3 - 6ax^2 -$
    $(6c + 12ab)x + (2ac - 8ab^2 - 4bc)] / (x^2 + 2bx + c)^3$.

33. Combine the sum rule, the power rule, and the quotient rule to get

$h'(r) = 13r^{12} - 4\sqrt{2}r^3 - [(1)(r^2 + 3) - (r)(2r)]/(r^2 + 3)^2 = 13r^{12} - 4\sqrt{2}r^3 - (-r^2 + 3)/(r^2 + 3)^2$; $h''(r) = 156r^{11} - 12\sqrt{2}r^2 - [(-2r)(r^2 + 3)^2 - (-r^2 + 3)(2)(r^2 - 3)(2r)]/(r^2 + 3)^4 = 156r^{11} - 12\sqrt{2}r^2 - [(-2r)(r^2 + 3) - (-r^2 + 3)(4r)]/(r^2 + 3)^3 = 156r^{11} - 12\sqrt{2}r^2 - (2r^3 - 18r)/(r^2 + 3)^3$.

37. Apply the power of a function rule with the product rule to get

$h'(x) = 4(x - 2)^3(x^2 + 2) + (x - 2)^4(2x) = 2(x - 2)^3(3x^2 - 2x + 4)$;

$h''(x) = 6(x - 2)^2(3x^2 - 2x + 4) + 2(x - 2)^3(6x - 2) = 2(x - 2)^2 \times (15x^2 - 20x + 16)$.

41. Mathematically, the first statement says: $dV/dt = kS$, where V is the volume, k is the proportionality constant and S is the surface area, $4\pi r^2$. By the chain rule, $dV/dt = (dV/dr)(dr/dt) = kS$. Since $V = 4\pi r^3/3$, we have $4\pi r^2(dr/dt) = k(4\pi r^2)$, which simplifies to $dr/dt = k$.

45. Denote the length of the legs by a and b , so the perimeter is $P = a + b + \sqrt{a^2 + b^2}$. Differentiate with respect to time: $dP/dt = da/dt + db/dt + (1/2)(a^2 + b^2)^{-1/2}(2da/dt + 2db/dt)$. At the moment in question, $a = b$, and $10^{-6} = (1/2)ab$, so $a = \sqrt{2} \cdot 10^{-3} = b$. Also, since the area is constant, one leg is decreasing its length while the other increases. Thus, $dP/dt = -10^{-4} + 10^{-4} + (1/2)(4 \cdot 10^{-6})^{-1/2} \times [(2)(-10^{-4}) + (2)(10^{-4})] = 0$.

49. Using the figure, we have $A = (\text{side})^2 - (1/2)(\text{base})(\text{height}) = (5x)^2 - (1/2)(2x)(3x) = 22x^2$. Therefore, $dA/dx = 44x$ and $d^2A/dx^2 = 44$.

53. Using the Pythagorean theorem, the hypotenuse of the triangle has
    length $13\sqrt{x}$. Then, the perimeter is $4(5x) + 3x + 2x + \sqrt{13}x =$
    $25x + \sqrt{13}x$. Solve for x and substitute into A from Exercise 49.
    $x = P/(25 + \sqrt{13})$ implies $A = 22P^2/(25 + \sqrt{13})^2$. Therefore,
    $dA/dP = 44P/(25 + \sqrt{13})^2$ and $dP/dx = 15 + \sqrt{13}$.

57. (a) Marginal cost is defined as $dC/dx = [5 - (0.02)x]$ dollars/case.

    (b) $(dC/dx)\big|_{84} = 5 - (0.02)(84) = \$3.32$.

    (c) According to part (a), marginal cost is a linear function with
        slope -0.02 , a decreasing function of x .

    (d) It is unreasonable for total cost to be less than or equal to
        zero. The quadratic formula, applied to $C(x)$, results in
        $x \approx 503.97$; therefore, when $x \geq 504$, the formula cannot
        be applicable.

61. The quotient rule gives $f'(x) = [(3x^2)(x^3 + 11) - (x^3 - 7)(3x^2)] /$
    $(x^3 + 11)^2 = 54x^2/(x^3 + 11)^2$, which is $54(2)^2/((2)^3 + 11)^2 =$
    $216/361$ at $x_0 = 2$. Thus, the tangent line is $y = y_0 +$
    $f'(x_0)(x - x_0) = 1/19 + (216/361)(x - 2)$.

65. The tangent line is given by $y = y(2) + (dy/dx)\big|_{t=2}(x - x(2))$,
    where $dy/dx = (dy/dt)/(dx/dt)$. $y(2) = 1 + \sqrt[3]{2} + 2 = 3 + \sqrt[3]{2}$;
    $x(2) = \sqrt{2} + 4 + 1/2 = 9/2 + \sqrt{2}$; $(dy/dt)/(dx/dt) = [(1/3)t^{-2/3} + 1] /$
    $[(1/2)t^{-1/2} + 2t - t^{-2}]$, and at $t = 2$, $dy/dx = [1/3(4)^{1/3} + 1] /$
    $[1/2\sqrt{2} + 15/4]$. Therefore, the tangent line is $y = (3 + \sqrt[3]{2}) +$
    $[(1 + 3\sqrt[3]{4}) \, 4\sqrt{2}/3 \, \sqrt[3]{4}(15\sqrt{2} + 2)](x - 9/2 - \sqrt{2})$.

69. (a) The linear approximation is given by $f(x_0 + \Delta x) \approx f(x_0) + f'(x_0)\Delta x$.

$f'(x) = [(40x^{39})(x^{29} + 1) - (x^{40} - 1)(29x^{28})]/(x^{29} + 1)^2$ and

$f'(1) = [(40)(2) - (0)(29)]/(2)^2 = 20$. Also, $f(1) = 0/2 = 0$.

Therefore, the linear approximation to $(x^{40} - 1)/(x^{29} + 1)$ at

$x_0 = 1$ is $20\Delta x$.

(b) x_0 and the function are the same as in part (a) . $\Delta x = 0.021$,

so the approximate value is $20(0.021) = 0.42$.

73. Applying the power of a function rule once gives the derivative as

$n[f(x)^m]^{n-1}(d/dx)[f(x)^m]$. Applying the rule again gives

$n[f(x)^m]^{n-1}m[f(x)^{m-1}]f'(x) = nm[f(x)^{mn-m+m-1}]f'(x) = nm[f(x)^{mn-1}]f'(x)$.

Applying the rule to the right-hand side gives $mn[f(x)^{mn-1}]f'(x)$,

which is the same.

77. The polynomial rule for antidifferentiation gives $\int(4x^3 + 3x^2 + 2x + 1)dx =$

$4x^4/4 + 3x^3/3 + 2x^2/2 + x + C = x^4 + x^3 + x^2 + x + C$.

81. This simplifies to $\int(-x^{-2} - 2x^{-3} - 3x^{-4} - 4x^{-5})dx$. The sum and power

rules for antidifferentiation may now be applied to get $-x^{-1}/(-1) -$

$2x^{-2}/(-2) - 3x^{-3}/(-3) - 4x^{-4}/(-4) + C = 1/x + 1/x^2 + 1/x^3 + 1/x^4 + C$.

85. The sum and power rules for antidifferentiation gives $\int(x^{3/2} + x^{-1/2})dx =$

$x^{5/2}(5/2) + x^{1/2}(1/2) + C = 2x^{5/2}/5 + 2\sqrt{x} + C$.

89. Use the formula $\int(ax + b)^n dx = (ax + b)^{n+1}/a(n+1) + C$ to get

$\int\sqrt{x - 1}dx = (x - 1)^{3/2}/(3/2) + C = 2(x - 1)^{3/2}/3 + C$.

93. Apply the sum rule for antiderivatives along with the formula

$\int(ax + b)^n dx = (ax + b)^{n+1}/a(n + 1) + C$. This gives $\int[(x - 1)^{1/2} -$

$(x - 2)^{5/2}]dx = (x - 1)^{3/2}/(3/2) - (x - 2)^{7/2}/(7/2) + C = 2(x - 1)^{3/2}/3 -$

$2(x - 2)^{7/2}/7 + C$.

97. The use of the chain rule gives $f'(x) = (1/2)x^{-1/2} - (1/2)[(x - 1)/$

$(x + 1)]^{-1/2}[(x + 1) - (x - 1)]/(x + 1)^2 = 1/2\sqrt{x} - (1/2)[(x + 1)/$

$(x - 1)]^{1/2}(2)/(x + 1)^2 = 1/2\sqrt{x} - 1/(x + 1)^{3/2}\sqrt{x - 1}$. By the definition

of antiderivatives, $\int [1/2\sqrt{x} - 1/(x + 1)^{3/2}\sqrt{x - 1}]dx =$

$\sqrt{x} - \sqrt{(x - 1)/(x + 1)} + C$.

101. The use of the chain rule gives $f'(x) = (1/2)[(x^2 + 1)/(x^2 - 1)]^{-1/2} \times$

$[(2x)(x^2 - 1) - (x^2 + 1)(2x)]/(x^2 - 1)^2 = (1/2)[(x^2 - 1)/(x^2 + 1)]^{1/2}(-4x)/$

$(x^2 - 1)^2 = -2x/(x^2 - 1)^{3/2}\sqrt{x^2 + 1}$. By the definition of antiderivatives,

$\int [-2x/(x^2 - 1)^{3/2}\sqrt{x^2 - 1}]dx = [(x^2 + 1)/(x^2 - 1)]^{1/2} + C$.

105. By the chain rule, we have $dD/dt = (dD/dv)\cdot(dv/dt) = 7(12) = 84$ pounds/second.

109. The proof is by induction on k . For $k = 1$, if $r'(x) = 0$, then r is

constant by Review Exercise 108. Hence, it is a polynomial.

Suppose the statement is true for $k - 1$. If $0 = (d^k/dx^k)(r(x)) =$
$(d^{k-1}/dx^{k-1})(r'(x))$, then $r'(x)$ is a polynomial by the induction

hypothesis. Let $g(x)$ be a polynomial such that $g'(x) = r'(x)$ (by the

antiderivative rule for polynomials). Then $(g - r)'(x) = 0$, so $g - r$

is constant, and hence, r is a polynomial.

TEST FOR CHAPTER 2

1. True or false:

(a) If $f'(x)$ exists, then $f''(x)$ also exists.

(b) The parametric equations $x = t^3$ and $y = t^3 + 8$ describe a
straight line.

(c) If $y = 3x^2 + 2x$, then $d^2y/dx^2 = (6x + 2)^2$.

(d) For differentiable functions f and g , the second derivative
of $f + g$ is $f'' + g''$.

(e) The curve described by $x = t^3 - 3$ and $y = -2t^3 + 2$ has a
constant slope.

2. Find dy/dx in each case:

(a) $xy^2 = 2y/x + 2$

(b) $3xy - \sqrt{xy} = y + x$

(c) $x(y + 3) + y/x - y^3 = y(x + 3)$

3. In each case, find dy/dx in terms of t :

(a) $x = 9t^3 + 8$, $y = 7t^2 - 8$

(b) $x = t^{3/2}$, $y = 8$

(c) $y = (t + 4)^2$, $x = \sqrt{t^2 - 3}$

4. Find a general formula for the second derivative of f/g . Assume
    f , f' , g , and g' are differentiable and $g \neq 0$.

5. Let $F(x)$ be a cubic function. If $F(-1) = 3$, $F'(0) = 3$, $F''(1) = 3$,
    and $F'''(3) = 3$, what is $F(x)$?

6. Differentiate $((1 + y^2)^3 + 1)^{-1/2}$ with respect to y .

7. Compute the second derivatives of the following functions:

(a) $f(x) = x^6 - 5x^3 + 3$

(b) $f(x) = -x + 6$

(c) $f(x) = (x + 4)(x^2 + 2)$

(d) $f(x) = 1/(3 - x)$

8. Suppose a square's side is increasing by 5 cm/sec. How fast is the
    area increasing when the length of a side is 10 cm. ?

9. $y = (x^{3/2} + 1)^{3/4}$ is a particle's position. What is the acceleration
    at time x ?

10. You're at the top of a 26 m ladder painting an office building. The
    other end of the ladder is being held by your partner who is 10 m from
    the building. At precisely 3 PM, your partner runs off for his coffee
    break leaving you to fall with the ladder at 50m/min. How fast is the
    other end of the ladder moving when you are halfway to breaking your bones?

ANSWERS TO CHAPTER TEST

1. (a) False; suppose $y = x^2$ if $x \leq 0$ and $y = 0$ if $x \geq 0$.

    (b) True

    (c) False; $d^2y/dx^2 = 6$.

    (d) True

    (e) True

2. (a) $(x^2y^2 + 2y)/(2x - 2x^3y)$

    (b) $(6y\sqrt{xy} - 2\sqrt{xy} - y)/(-6x\sqrt{xy} + 2\sqrt{xy} + x)$

    (c) $(3x^2 - y)/(3y^2x^2 - x + 3x^2)$

3. (a) $14/27t$

    (b) 0

    (c) $2(t + 4)\sqrt{t^2 - 3}/t$

4. $[g(f''g - g''f) - 2g'(f'g - g'f)]/g^3$

5. $F(x) = x^3/2 + 3x + 13/2$

6. $-3y(1 + y^2)^2/((1 + y^2)^3 + 1)^{3/2}$

7. (a) $30x^4 - 30x$

    (b) 0

    (c) $6x + 8$

    (d) $2/(3 - x)^3$

8. $100cm^2/sec$

9. $-9[3/4(x^{3/2} + 1) - 1/x^{3/2}]x/16(x^{3/2} + 1)^{1/4}$

10. $600/\sqrt{532}$ m/min

CHAPTER 3

GRAPHING AND MAXIMUM-MINIMUM PROBLEMS

3.1 Continuity and the Intermediate Value Theorem

PREREQUISITES

1. Recall the definition of continuity (Section 1.2).

2. Recall how to compute limits (Section 1.2).

PREREQUISITE QUIZ

1. What two conditions must be met for a function f to be continuous

    at x_0 ?

2. Find the following limits:

    (a) $\lim\limits_{x \to 2} x^3$

    (b) $\lim\limits_{x \to 0} [(x^2 + 5x)/x]$

    (c) $\lim\limits_{x \to 4} \sqrt{x^2 + 9}$

GOALS

1. Be able to describe the relationship between continuity and

    differentiability.

2. Be able to explain the intermediate value theorem in laymen's terms.

STUDY HINTS

1. Points of continuity. By studying Example 1, you should notice that a function may be continuous at a point or on an interval which is composed of a set of points, at each of which the function is continuous.

2. Continuity at endpoints. The definition of continuity at a point given on p. 63 in the text needs a slight modification at endpoints. The reason is that two-sided limits do not exist at endpoints. This problem is dealt with by using one-sided limits when speaking about continuity at endpoints. (See p. 65).

3. Continuity tests. Differentiability at a point implies continuity at that same point. Rational functions are differentiable and therefore, are continuous, except where the denominator becomes zero. Continuity alone does not imply differentiability.

4. Intermediate value theorems. Notice that both versions require continuity. The first version simply says that to get from one side of $y = c$ to the other side, one must cross it. The second version says that if one doesn't cross $y = c$, one stays on the side one started on.

5. Method of bisection. Study Example 7 well; the technique will be very useful in the future for computing numerical answers. The interval does not always have to be halved; it is only a convenient way to converge upon the answer.

SOLUTIONS TO EVERY OTHER EXERCISE

1. (a) The function jumps at $x_0 = \pm 1$, the points of discontinuity. Thus, it is continuous on $(-\infty, -1)$, $(-1, 1)$, and $(1, +\infty)$.

1. (b) Even though the function is not differentiable at $x_0 = 2$,

$\lim\limits_{x \to 2} f(x) = f(2)$,   so the function is continuous   for all x .

(c) The limit does not exist at $x_0 = -2$,   so the function is

continuous on $(-\infty, -1)$ and $(-1, +\infty)$.

5. At $x_0 = 1$,   the denominator is $4 \neq 0$,   so the function is con-

tinuous by the rational function rule.

9. $f(x)$ vanishes only at $x = \pm 1$.   Thus, by the rational function rule,

$f(x)$ is continuous on $(-\infty, -1)$, $(-1, 1)$ and $(1, \infty)$,   and in

particular, $[-1/2, 1/2]$.

13. This exercise is analogous to Example 4. We must show that

$\lim\limits_{x \to x_0} (f + g)(x) = (f + g)(x_0)$.   By the sum rule for limits,

$\lim\limits_{x \to x_0} [f(x) + g(x)] = \lim\limits_{x \to x_0} f(x) + \lim\limits_{x \to x_0} g(x) = f(x_0) + g(x_0)$,   since

f and g are continuous at x_0 .   In addition, $(f + g)(x_0) =$

$f(x_0) + g(x_0)$,   so $\lim\limits_{x \to x_0} (f + g)(x) = (f + g)(x_0)$ as required.

17. We must show that there exists s such that $f(s) = -s^5 + s^2 - 2s + 6 = 0$.

Now $f(2) = -26 < 0$ and $f(-2) = 46 > 0$,   so by the intermediate value

theorem (first version), we conclude that there is some s_0 in $(-2, 2)$

satisfying $f(s_0) = 0$.   (Other intervals also work, for example, $[0, 2]$.)

21. $f(-3) = -28$, $f(0) = 2$, $f(1.3) = -0.093$, and $f(2) = 2$.   Combining

this information with the fact that $f(\sqrt{2}) = f(-\sqrt{2}) = f(1) = 0$ and the

second version of the intermediate value theorem, we conclude that $f(x)$

is negative on $(-\infty, -\sqrt{2})$ and $(1, \sqrt{2})$. $f(x)$ is positive on $(-\sqrt{2}, 1)$

and $(\sqrt{2}, +\infty)$.

25. By the intermediate value theorem (second version), we know that

$f(x) - 2 < 0$ on $[-1, 1]$ because $f(0) - 2 = -2$.   From $f(x) - 2 < 0$,

we conclude $f(x) < 2$ on $[-1, 1]$.

29. In order for $f(x)$ to be continuous, we must have $\lim\limits_{x\to 2} f(x) = f(2)$.

$\lim\limits_{x\to 2}[(x^2 - 4)/(x - 2)] = \lim\limits_{x\to 2} (x + 2) = 4$, so define $f(2)$ as 4 .

33. In order for $f(x)$ to be continuous, we must have $f(1) = \lim\limits_{x\to 1} f(x)$.

For $x \neq 1$, $f(x) = (x^2 - 1)/(x - 1) = [(x - 1)(x + 1)]/(x - 1) =$

$x + 1$. Since $\lim\limits_{x\to 1} f(x) = \lim\limits_{x\to 1} (x + 1) = 2$, we define $f(1)$ as 2 .

37. The function $f(x) = 1/(x - 1)$ is not continuous on the interval

$[0,2]$, since it is not defined at $x = 1$; therefore, the inter-

mediate value theorem does not apply to f and there is no

contradiction.

41. Let x_0 be a root of $f(x) = x^5 + x^2 + 1$. The method of bisection,

discussed in Example 7, is employed to develop the following chart.

Midpoint of I	Nature of f(midpoint)	Conclude that x_0 lies in I
	$f(-2) < 0 < f(-1)$	$(-2, -1)$
-1.5	$f(-1.5) < 0 < f(-1)$	$(-1.5, -1)$
-1.25	$f(-1.25) < 0 < f(-1)$	$(-1.25, -1)$
-1.125	$f(-1.125) > 0 > f(-1.25)$	$(-1.25, -1.125)$
-1.1875	$f(-1.1875) > 0 > f(-1.25)$	$(-1.25, -1.1875)$
-1.21875	$f(-1.21875) < 0 < f(-1.1875)$	$(-1.21875, -1.1875)$
-1.203125	$f(-1.203125) < 0 < f(-1.1875)$	$(-1.203125, -1.1875)$
-1.1953125	$f(-1.1953125) < 0 < f(-1.1875)$	$(-1.1953125, -1.1875)$
-1.1914062	$f(-1.1914062) > 0 > f(-1.1953125)$	$(-1.1953125, -1.1914062)$
-1.1933593	$f(-1.1933593) > 0 > f(-1.1953125)$	$(-1.1953125, -1.1933593)$
-1.1943359	$f(-1.1943359) < 0 < f(-1.1933593)$	$(-1.1943359, -1.1933593)$

So -1.194 approximates x_0 to within 0.001 .

45. It is simplest to describe an example. Consider approximating the root

$\sqrt{7}$ of $f(x) = x^2 - 7$ to within 0.01 . Method 1 (bisection method):

division point is the midpoint (see the solution to Exercise 24). Method

2: division point is the x-intercept $\bar{x}$ of the line L through $(a,f(a))$

and $(b,f(b))$ where $\bar{x}$ approximates the root and the nature of $f(\bar{x})$

determines the next interval (a,b) in which the root can be found.

45 (continued).

Start with $(a,b) = (2,3)$ as in Exercise 24. $f(2)$ and $f(3)$ have

opposite signs, so Method 2 can be used. $\sqrt{7} \approx 2.6457513$.

n	$(a,f(a))$	$(b,f(b))$	L	$\bar{x}$	Accuracy	Nature of $f(\bar{x})$	(a,b)
1	$(2,-3)$	$(3,2)$	$y = 5x - 13$	2.6	0.1	$f(\bar{x}) < 0 < f(3)$	$(2.6,3)$
2	$(2.6,-0.24)$	$(3,2)$	$y = (5.6)x - 14.8$	2.6428571	0.01	$f(\bar{x}) < 0 < f(3)$	$(2.6428571,3)$

Hence we see that Method 2 requires only $n = 2$ steps while Method 1

requires $n = 7$ steps to attain the same accuracy. Now consider the

method of dividing the interval(s), at each step, into 10 (instead of

2) equal parts and looking for sign changes at the 9 interior division

points (instead of the midpoint). At the nth step, there would be 10^n

subintervals and so the size of the interval in which a solution can be

found would be $1/10^n$ (assuming that the original interval has unit

length). Hence n , the number of steps required to achieve the

accuracy A , satisfies $1/10^n \leqslant A$. By letting A take the values

$1/10$, $1/100$, ... , $1/10^n$, ... , we see that the number of steps

required coincides with the number of digits of accuracy desired. Hence

this method is more appropriate to the decimal system, but not necessarily

the more efficient.

49. Let $p(x) = b_n x^n + \ldots + b_0$ with $n = $ odd, $b_n \neq 0$. Then $f(x) = $
$p(x)/b_n = x^n + a_{n-1}x^{n-1} + \ldots + a_0 (a_i = b_i/b_n$, $i = 0$, ... , $n - 1)$
has the same roots as $p(x)$. Hence it suffices to show that f has a
root. $f(x)/x^n = 1 + a_{n-1}/x + \ldots + a_0/x^n \geqslant 1 - (1/|x|)\{|a_{n-1}| +$
$|a_{n-2}|/|x| + \ldots + |a_0|/|x|^{n-1}\}$. For $|x| > 1$ and $|x| > 2\{|a_0| + \ldots +$
$|a_{n-1}|\}$, $1/|x| < 1$ and $|x|/2 > \{|a_0| + \ldots + |a_{n-1}|\}$. So
$(1/|x|)\{|a_{n-1}| + |a_{n-2}|/|x| + \ldots + |a_0|/|x|^{n-1}\} < (1/|x|)\{|a_{n-1}| + \ldots +$
$|a_0|\} < (1/|x|)(|x|/2) = 1/2$. Hence $f(x)/x^n \geqslant 1 - (1/|x|)\{|a_{n-1}| +$

49 (continued).

$|a_{n-2}|/|x| + \ldots + |a_0|/|x|^{n-1}\} > 1 - 1/2 > 0$. So $f(x)$ and x^n

have the same sign, i.e., $f(x) < 0$ if x is large negative and

$f(x) > 0$ if x is large positive since n is odd. By the intermediate

value theorem, there is an x_0 with $f(x_0) = 0$.

SECTION QUIZ

1. (a) Is it true that all continuous functions are differentiable?

    (b) Is it true that all differentiable functions are continuous?

2. $f(x) = x^3 + 0.9x^2 - 0.75x - 0.25$ has three roots. Use the method

    of bisection or a similar technique to find them within 0.01 .

3. Let $g(x) = (x^2 - 4x - 5)/(x + 1)$. This is not defined at $x = -1$.

    How should $g(-1)$ be defined to maintain continuity?

4. True or false: If $f(-1) = -1$ and $f(1) = 1$, then $f(x)$ must be

    zero for some x in $(-1,1)$. Explain your answer.

5. Being employed by a maharajah can have its interesting moments. For

    example, just yesterday morning, one of the servants took one of the

    pet elephants out for a ride. Their position was given by $y = 5x$,

    until the elephant saw a mouse. Suddenly, at $x = 10$, their position

    became $y = -250 + 30x$.

    (a) Find the velocity function.

    (b) Is the position function continuous?

    (c) Is the velocity function continuous?

    (d) Does the intermediate value theorem tell you that somewhere in

         $(5,15)$, the velocity must be 20, which is between 5 and 30.

         Explain.

SOLUTIONS TO PREREQUISITE QUIZ

1. $f(x_0)$ exists and $\displaystyle\lim_{x \to x_0} f(x) = f(x_0)$.

2. (a) 8

   (b) 5

   (c) 5

SOLUTIONS TO SECTION QUIZ

1. (a) No; consider $|x|$ when $x = 0$.

   (b) Yes

2. -1.32 , -0.27 , 0.70

3. -6

4. True if f is continuous; false if f is not continuous.

5. (a) $v = \begin{cases} 5 & \text{if } x < 10 \\ 30 & \text{if } x > 10 \end{cases}$

   (b) Yes

   (c) No

   (d) No; the velocity function is not continuous, so the intermediate value theorem does not apply.

3.2   Increasing and Decreasing Functions

PREREQUISITES

1.   Recall how to solve inequalities (Section R.1).

2.   Recall the intermediate value theorem (Section 3.1).

3.   Recall how to differentiate rational functions (Sections 1.4 and 1.5).

PREREQUISITE QUIZ

1.   Find the solution set of $x^2 - 7x + 10 \geqslant 0$.

2.   If $f(-2) = 3$ and $f(1) = -2$, and f is continuous, is it true that $f(x_0) = 0$ for some x_0 such that $-2 \leqslant x_0 \leqslant 1$?   Cite a theorem which defends your answer.

3.   Differentiate the following functions:

(a) $2x^5 - 3x^2 + 2$

(b) $(x + 8)(x^3 + 3x^2 + 2)$

(c) $(x - 3)/(x^2 + 5x)$

GOALS

1.   Be able to determine whether a function is increasing or decreasing by using the derivative.

2.   Be able to classify critical points as a local minimum, a local maximum, or neither.

STUDY HINTS

1.   <u>Definition of increasing.</u> Fig. 3.2.2 shows the graph of an increasing function which is also changing signs. By looking at the graph, the definition of increasing should become intuitively obvious. As you move from left to right, x is increasing and f(x) is increasing

1 (continued).

("getting larger") as well in a small interval (a,b) .

Note that the definition of increasing may not apply if the interval is chosen to be too large. For example, if we chose $[a, x_3]$ in
Fig. 3.2.4, the function decreases at the right.

2.   Changing signs. It should be intuitively obvious that an increasing
function passing through $y = 0$ is changing from negative to positive.
(See Fig. 3.2.2.) Similar reasoning applies to decreasing functions
passsing through $y = 0$. (See Fig. 3.2.3.)

3.   Increasing-decreasing test. Know this test well! A positive derivative means f is increasing; a negative derivative means f is decreasing. A zero derivative means that the slope of the graph is flat.
Test you understanding by explaining all of the answers in Example 8.

4.   Increasing on intervals. One can conclude that a function is increasing
on an entire interval provided the derivative is positive throughout that
interval. A similar statement may be made for the decreasing case. By
using the intermediate value theorem, one can conclude that increasing
and decreasing intervals are separated by points where $f'(x) = 0$.

5.   Critical point test. The test is very important to know. The example
$y = x^3$ is a common one to use when asked for a counterexample to demonstrate when the test fails. For $y = x^3$, $x_0 = 0$ is a critical point,
but zero is neither a minimum nor a maximum. Also, note that the test
only applies to differentiable functions. Think about $f(x) = |x|$.

6.   First derivative test. This test should be understood, not memorized.
Think about what the signs of the derivative mean and be able to reproduce Fig. 3.2.17.

SOLUTIONS TO EVERY OTHER ODD EXERCISE

1.   Use the definitions of increasing and decreasing given at the beginning
of this section. From the graph, we see that (a,b) may be chosen to

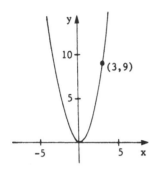

be $(2,4)$. If $2 < x < 3$, then $f(x) < 9 = f(x_0)$. If $3 < x < 4$, then $f(x) > 9 = f(x_0)$. Thus, by definition, $f(x)$ is increasing at $x_0 = 3$.

5.   For the general case, the chosen interval (a,b) must not be too large.
In the case, (a,b) may be chosen as large as desired; we choose $(0,1)$.
For $f(x) = 2x - 1$, $f(0) = -1 < 0$, $f(1/2) = 0$, and $f(1) = 1 > 0$,
so f changes from negative to positive at $x_0 = 1/2$.

9.   Using the increasing-decreasing test, we get $f'(x) = 3x^2 + 1$ and
$f'(0) = 1$. Since $f'(x_0)$ is positive, f is increasing at $x_0 = 0$.

13. Here, $f'(t) = 5t^4 - 4t^3 + 4t$ and $f'(1) = 5$ is positive. Thus, $f(t)$
is increasing at $t = 1$, i.e., the particle is moving to the right.

17. Here, $f'(x) = 2x$, so $f'(x) > 0$ if $x > 0$ and $f'(x) < 0$ if $x < 0$.
Therefore, the increasing-decreasing test states that f is increasing
on $(0, \infty)$ and decreasing on $(-\infty, 0)$.

21. The derivative in (e) is a positive constant, so the function is an
increasing linear function, namely, (5). The derivative in (b) is always
positive, so the function is always increasing. Of the remaining functions,
only (4) has this property. The derivative in (c) indicates that the
function near $x = 0$ is increasing for $x < 0$ and decreasing for
$x > 0$. Thus, (c) must match with (1). For (d), we look for a function

21 (continued).

which decreases for all x < 0 and increases for all x > 0 . Only
(2) satisfies this. Note that (3) does not always increase for x > 0 .
It matches (a). Thus, the answer is (a) - (3) , (b) - (4) , (c) - (1) ,
(d) - (2) , (e) - (5) .

25. We use the definitions of local minimum and local maximum on p. 151.
The local minimum points occur at x_1 , x_3 , and x_5 . Local maximum
points are at x_2 and x_4 . x_6 is neither a local minimum nor a local
maximum point.

29. For $f(x) = x^3 + x^2 - 2$, $f'(x) = 3x^2 + 2x = x(3x + 2)$. $f'(x) = 0$
at $x = 0$ and $x = -2/3$, so these are the critical points. The
sign of f' changes from negative to positive at $x = 0$ and from positive to negative at $x = -2/3$. Hence, the first derivative test states
that 0 is a local minimum and -2/3 is a local maximum.

33. $\ell'(r) = 2(r^4 - r^2)(4r^3 - 2r) = 4r^3(r^2 - 1)(2r^2 - 1)$, so the critical
points are 0 , ± 1 , $\pm 1/\sqrt{2}$. $\ell'(r)$ changes from negative to positive
at $r = -1,0,1$. Therefore, these are local minima, by the first
derivative test. $\ell'(r)$ changes from positive to negative at $\pm 1/\sqrt{2}$,
the local maxima.

37. $f'(x) = m$. For $m = -2$, the sign change is from positive to negative
since the function is decreasing. There is no sign change when $m = 0$.
For $m = 2$, the sign change is from negative to positive since the
function is increasing.

41. $f(x) = x^2 - 4x + 4$ has a double root at $x = 2$, so $f(2) = 0$. $f'(x) = 2x - 4$, which is negative for $x < 2$ and positive for $x > 2$. Thus,
$f'(x)$ changes sign at 2 ; therefore, $x = 2$ is a local minimum. Conse-
quently, f does not change sign anywhere.

45. For $f(x) = 2x^3 - 5x + 7$, $f'(x) = 6x^2 - 5$. $f'(x)$ is positive for $|x| > \sqrt{5/6}$ and negative for $|x| < \sqrt{5/6}$. Hence f is increasing on $(-\infty, -\sqrt{5/6})$ and $(\sqrt{5/6}, \infty)$; f is decreasing on $(-\sqrt{5/6}, \sqrt{5/6})$.

49. $f(x) = ax^2 + bx + c$, so $f'(x) = 2ax + b$. The hypothesis that $f' < 0$ for $x < 2$ and $f' > 0$ for $x > 2$ implies $x = 2$ is a local minimum and hence a critical point, i.e., $0 = f'(2) = 4a + b$. The hypothesis that $x = 1$ is a zero of f implies $0 = f(1) = a + b + c$. Hence $b = -4a$, $c = -a - b = 3a$, and so $f(x) = ax^2 - 4ax + 3a = a(x^2 - 4x + 3)$ where $a > 0$ since a parabola with a minimum is a parabola which opens upward.

53. An increasing f and g at x_0 implies that $f(x) < f(x_0) < f(y)$ and $g(x) < g(x_0) < g(y)$ for $x < x_0 < y$. Hence, $f(x) + g(x) < f(x_0) + g(x_0) < f(y) + g(y)$ for $x < x_0 < y$, showing that $f + g$ is increasing at x_0.

57. (a) gh is increasing if $(gh)' = g'h + gh' > 0$. Divide both sides of the inequality by gh (which is positive and hence preserves the inequality) to yield the desired criterion $g'/g + h'/h > 0$. gh is decreasing if $g'/g + h'/h < 0$.

    (b) g/h is increasing if $(g/h)' = (g'h - h'g)/h^2 > 0$. Multiplying through by (h/g) gives $g'/g - h'/h > 0$. g/h is decreasing if $g'/g - h'/h < 0$.

61. Since the graph does not change sign at $x = 0$, the quartic has the form $p(x) = cx^2(x + b)(x - b) = cx^4 - cb^2x^2$, where c is a non-zero constant. Thus, $p'(x) = 4cx^3 - 2cb^2x$, which is 0 at $x = 0, \pm a$. When $x = a$, $4ca^3 - 2cb^2a = 0$ or $2a^2 = b^2$. We obtain the same relationship when $x = -a$. At $x = b$, we need $p'(x) > 0$, so $c > 0$. Hence, the polynomial is $p(x) = cx^4 - cb^2x^2$, $c > 0$, and $2a^2 = $

SECTION QUIZ

1. Consider the function $f(x) = x^3/(1 - x^2)$. On what intervals is f

    increasing? decreasing?

2. What are the critical points of $x^3/(1 - x^2)$? Classify any critical

    points as a local minimum, local maximum, or neither.

3. Sketch examples of the following types of critical points:

    (a) local minimum

    (b) local maximum

    (c) neither a local minimum nor a local maximum.

4. True or false: All local minima and local maxima are points where

    the derivative is zero.

5. Your house guest snores every evening with the same pattern. In hours

    from midnight, the volume is given by $S(t) = t^3 - 9t^2 + 15t$,

    $0 \leqslant t \leqslant 6$. If $S(t) < 0$, your house guest is not snoring.

    (a) What are the critical points of $S(t)$?

    (b) Classify the critical points as minima or maxima.

    (c) Where is $S(t)$ increasing? decreasing?

SOLUTIONS TO PREREQUISITE QUIZ

1. $x \leqslant 2$ or $x \geqslant 5$

2. Yes, use the intermediate value theorem.

3. (a) $10x^4 - 6x$

    (b) $(x^3 + 3x^2 + 2) + (x + 8)(3x^2 + 6x) = 4x^3 + 33x^2 + 48x + 2$

    (c) $(-x^2 + 6x + 15)/(x^2 + 5x)^2$

SOLUTIONS TO SECTION QUIZ

1. Increasing on $(-\sqrt{3},-1)$, $(-1,1)$, and $(1,\sqrt{3})$; decreasing on
    $(-\infty,-\sqrt{3})$, $(\sqrt{3}, \infty)$.

2. $x = -\sqrt{3}$: local minimum; $x = 0$: neither; $x = \sqrt{3}$: local maximum.

3. (a)   (b)

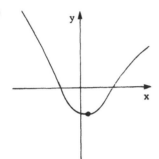   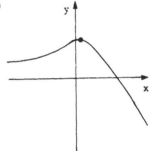

    (c)

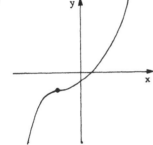

4. False; consider $f(x) = |x|$ at $x = 0$.

5. (a) $t = 1,5$

    (b) $t = 1$: maximum; $t = 5$: minimum.

    (c) Increasing on $(0,1)$, $(5,6)$; decreasing on $(1,5)$.

3.3   The Second Derivative and Concavity

PREREQUISITES

1.   Recall how derivatives can be used to show that a function is increasing
     or decreasing (Section 3.2).

2.   Recall how to compute a second derivative (Section 2.1).

PREREQUISITE QUIZ

1.   Let $g(t) = t^3 + 3t^2$. On what intervals is $g(t)$ increasing?

2.   Let $f(x) = x/(x + 2)$. On what intervals is $f(x)$ increasing? Decreasing?

3.   Find the second derivatives of the functions given in Questions 1 and 2.

GOALS

1.   Be able to use the second derivative to determine the concavity of a
     function.

2.   Be able to find the inflection points of a function.

STUDY HINTS

1.   Second derivative test.   There is a simple way to remember that $f''(x_0) > 0$
     and $f'(x_0) = 0$ implies a local minimum. Since $f'' > 0$, f' must be in-
     creasing. And since $f'(x_0) = 0$, $f'(x) < 0$ on the left of x_0 and
     $f'(x) > 0$ on the right side. Thus $f(x)$ is decreasing to a local mini-
     mum at x_0 , where it begins to increase again. Deriving the second
     derivative test in this fashion may be easier than memorizing it.

2.   Concavity.   A curve is concave upward if it can "hold water." In order
     for this to occur, the slopes must be getting larger, i.e., $f'(x)$ is
     increasing or $f''(x) \geqslant 0$. Similarly, $f''(x) \leqslant 0$ implies downward con-
     cavity, and these curves "spill water."

3. Concave up on intervals. As with the concept of increasing and de-
    creasing, an entire interval may possess a certain concavity. By
    using the intermediate value theorem, one can conclude that these
    intervals are separated by points x where $f''(x) = 0$.

4. Inflection points. Concavity changes at those points where $f''(x) = 0$
    and $f'''(x) \neq 0$. Notice the similarity of this definition with that
    of minimum and maximum points, where $f'(x)$ changes signs.

SOLUTIONS TO EVERY OTHER ODD EXERCISE

1. $f'(x) = 6x$, which is 0 at $x = 0$; $f''(x) = 6$ is positive, so by
    the second derivative test, $x = 0$ is a local minimum.

5. $f'(x) = 4x/(x^2 + 1)^2$, which is 0 at $x = 0$; $f''(x) = (4 - 12x^2)/$
    $(x^2 + 1)^3$. $f''(0) = 4$ which is positive, so by the second derivative
    test, $x = 0$ is a local minimum.

9. Since $f'(x) = 6x + 8$, $f''(x) = 6$ which is positive for all x . Thus,
    $f(x)$ is concave upward everywhere.

13. Since $f'(x) = -1/(x - 1)^2$, $f''(x) = 2/(x - 1)^3$. The second derivative
    is positive for $x > 1$ and negative for $x < 1$. Thus, $f(x)$ is con-
    cave upward on $(1,\infty)$ and concave downward on $(-\infty,1)$.

17. For $f(x) = x^3 - x$, $f'(x) = 3x^2 - 1$, $f''(x) = 6x$, and $f'''(x) = 6$.
    The second derivative vanishes at $x = 0$ and since $f'''(0) \neq 0$, it is
    an inflection point.

21. $f'(x) = 4(x - 1)^3$; $f''(x) = 12(x - 1)^2 = 0$ at $x = 1$. Note that
    $f''(x) \geqslant 0$ for all x , so it does not change sign; therefore, there
    are no inflection points.

25. (a) x_0 is a local maximum point since $f'(x)$ changes from positive
    to negative.

    (b) x_0 is an inflection point since $f(x)$ changes from concave
    downward to concave upward.

    (c) x_0 is neither. $f'(x)$ is positive on both sides of x_0 and
    $f''(x) = 0$ on the left of x_0 , so $f''(x)$ does not change sign.

    (d) x_0 is a local maximum point since $f'(x)$ changes from positive
    to negative.

    (e) x_0 is a local maximum point since $f'(x)$ changes from positive
    to negative.

    (f) x_0 is an inflection point since $f(x)$ changes from concave upward
    to concave downward.

    (g) x_0 is an inflection point since $f(x)$ changes from concave
    downward to concave upward.

    (h) x_0 is a local minimum point since $f'(x)$ changes from negative
    to positive.

29. $f(x) = x^3 + 2x^2 - 4x + 3/2$; $f'(x) = 3x^2 + 4x - 4 = 0$ at $x = 2/3$, -2 .
    f' is negative between its roots and positive outside its roots. Hence
    f is decreasing on $(-2, 2/3)$ and increasing on $(-\infty, -2)$ and $(2/3, \infty)$.
    $f''(x) = 6x + 4 = 0$ at $x = -2/3$; $f'''(-2/3) = 6 \neq 0$, so $x = -2/3$
    is an inflection point. As $f''(x) < 0$ for $x < -2/3$ and $f''(x) > 0$
    for $x > -2/3$, f is concave downward on $(-\infty, -2/3)$ and concave upward
    on $(-2/3, \infty)$.   In particular, $f''(-2) < 0$ and $f''(2/3) > 0$, so $x = -2$
    is a local maximum and $x = 2/3$ is a local minimum.

33. f has inflection points at 1 and 2 if $f''(x) = k(x - 1)(x - 2)$,

   $k \neq 0$. Practice with differentiation tells us that if f'' is a

   quadratic, then f' is a cubic and f is a quartic. Let $f(x) =$

   $ax^4 + bx^3 + cx^2 + dx + e$, then $f'(x) = 4ax^3 + 3bx^2 + 2cx + d$;

   $f''(x) = 12ax^2 + 6bx + 2c$ which should $= k(x^2 - 3x + 2)$. Hence

   $a = k/12$; $b = -k/2$; $c = k$; d , e are arbitrary. So $f(x) =$

   $kx^4/12 - kx^3/2 + kx^2 + dx + e$. A specific example is $f(x) =$

   $x^4/12 - x^3/2 + x^2 - x - 1$. (Alternatively, use the method of

   antidifferentiation in Section 2.5 .)

37. (a) For x near x_0 , the linear approximation of $f(x)$ at x_0 is

      the linear function $\ell(x) = f(x_0) + f'(x_0)(x - x_0)$. $f(x) =$

      $x^3 - x$; $f(-1) = f(0) = f(1) = 0$. $f'(x) = 3x^2 - 1$; $f'(\pm 1) =$

      2 ; $f'(0) = -1$. Hence $\ell(x)$ at $x_0 = -1$, 0 , 1 is $2(x + 1)$,

      $-x$, $2(x - 1)$, respectively.

   (b)

x_0	$f''(x_0)$	Δx	$x = x_0 + \Delta x$	$f(x)$	$\ell(x)$	$e(x) = f(x) - \ell(x)$ error
-1	-6	1	0	0	2	-2
		-1	-2	-6	-2	-4
		0.1	-0.9	0.171	0.2	-0.029
		-0.1	-1.1	-0.231	-0.2	-0.031
		0.01	-0.99	0.019701	0.02	-0.000299
		-0.01	-1.01	-0.020301	-0.02	-0.000301
0	0	1	1	0	-1	1
		-1	-1	0	1	-1
		0.1	0.1	-0.099	-0.1	0.001
		-0.1	-0.1	0.099	0.1	-0.001
		0.01	0.01	-0.009999	-0.01	0.000001
		-0.01	-0.01	0.009999	0.01	-0.000001
1	6	1	2	6	2	4
		-1	0	0	-2	2
		0.1	1.1	0.231	0.2	0.031
		-0.1	0.9	-0.171	-0.2	0.029
		0.01	1.01	0.020301	0.02	0.000301
		-0.01	0.99	-0.019701	-0.02	0.000299

37. (b)   (continued)

The table shows that $\dot{e}(x) < 0$ if $f''(x_0) < 0$ and $e(x) > 0$ if $f''(x_0) > 0$. If $f''(x_0) = 0$, $e(x)$ is comparatively smaller and decreases faster as Δx decreases. The sign of the error is the same as for Δx when $f''(x_0) = 0$.

41. $s'(t) = 40 - 32t$, which vanishes at $t = 5/4$; $s''(5/4) = -32 < 0$, so the maximum height is attained at $t = 5/4$ seconds. The maximum height is $s(5/4) = 28$ feet.

45. No. Counterexample: $f(x) = x^2 + 1 > 0$ for all x ; $f''(x) = 2 \neq 0$ for all x , so f has no inflection points. Let $g(x) = 1/f(x) = 1/(x^2 + 1)$; $g'(x) = -2x/(x^2 + 1)^2$; $g''(x) = 2(3x^2 - 1)/(x^2 + 1)^3 = 0$ at $x = \pm\sqrt{3}/3$. As $x^2 + 1 > 0$, $(x^2 + 1)^3 > 0$, so $g''(x)$ changes sign wherever $3x^2 - 1$ does. Since $3x^2 - 1$ is a quadratic with two distinct roots, it changes sign at its roots $x = \pm\sqrt{3}/3$. Hence g has inflection points at $x = \pm\sqrt{3}/3$ while f has none.

SECTION QUIZ

1. Consider the function $f(x) = x^3/(1 - x^2)$. On what intervals is $f(x)$ concave upward? concave downward?

2. Where is $f''(x) = 0$ for $f(x) = x^3/(1 - x^2)$? Is this an inflection point? Why or why not?

3. True of false: Suppose f is twice differentiable. If $x = 3/2$ is the only inflection point of a continuous function f and $f''(x) < 0$ for $x < 3/2$, then $f''(x) \geqslant 0$ for $x > 3/2$. Explain.

4. Paranoid Pete saw some UFO's last night. Thinking that we are about to be invaded, he orders an underground shelter to be built. In order to fool the invaders, he wants an uneven terrain. The shape of the terrain

4.   (continued)

   is described by $w = x^4/12 - x^3/5 + x^2/10 + 2x - 3$, $x \geqslant 0$. In

   addition, Paranoid Pete is very concerned about drainage in case of

   rain; therefore, he needs to know the concavity of the terrain.

   (a) Where are the inflection points, if any?

   (b) Discuss the concavity of the terrain.

ANSWERS TO PREREQUISITE QUIZ

1. $(-\infty,-2)$ and $(0,\infty)$

2.   Increasing for $x \neq -2$; decreasing nowhere.

3. $g''(t) = 6t + 6$; $f''(x) = -4/(x + 2)^3$

ANSWERS TO SECTION QUIZ

1.   Upward: $(-\infty,-1)$, $(0,1)$; downward: $(-1,0)$, $(1,\infty)$

2.   0 ; yes; $f''' \neq 0$

3.   True; concavity changes at an inflection point. $f'' < 0$ implies
   downward concavity, so the graph must be concave upward for $x > 3/2$.

4.   (a) $x = 1/5$, 1

   (b) Concave downward: $[0,1/5)$, $(1,\infty)$; concave upward: $(1/5,1)$

3.4   Drawing Graphs

PREREQUISITES

1. Recall how derivatives give information about local extrema, increasing, decreasing, concavity, and inflection points (Sections 3.2 and 3.3).

2. Recall how to compute limits at infinity (Section 1.2).

3. Recall how to compute infinite limits (Section 1.2).

PREREQUISITE QUIZ

1. If x_0 is a local extreme point, what can you say about the value of $f'(x_0)$?

2. If a function g is increasing at x_0 , what can you say about $g'(x_0)$?

3. If $g'(t_0) > 0$, is g increasing or decreasing at t_0 ?

4. If $f''(x_0) < 0$, is f concave upward or downward at x_0 ?

5. Compute the following limits, stating $\pm\infty$, if appropriate:

   (a) $\lim\limits_{x \to \infty} [(2x^2 - 1)/(x^2 + 1)]$

   (b) $\lim\limits_{x \to -\infty} [(3x^3 + x - 1)/(x^3 + x)]$

   (c) $\lim\limits_{x \to 0+} (1/x)$

   (d) $\lim\limits_{x \to 2-} [1/(x - 2)]$

GOALS

1. Be able to use calculus for the purpose of sketching graphs.

STUDY HINTS

1. Symmetries. Before jumping into the problem, note any symmetries. If any symmetry exists, your work is reduced by half. The two important symmetries are those about the y-axis and about the origin. Note that

1 (continued).

if all of the exponents in a polynomial are <u>even</u>, the function is <u>even</u>. Similarly, all odd exponents make the function odd. The last two statements apply only to polynomials. A quotient such as $(x^3 + x^5)/(x + x^3)$ has all odd exponents, but it is an even function. Remember that all constant terms have an even exponent, namely, zero.

2.   <u>Horizontal asymptotes</u>. If limits exist at $\pm\infty$, these limits are the horizontal asymptotes. Each function can have at most two horizontal asymptotes. Note that a function may cross the asymptote one or more times as x increases; consider $f(x) = x/(1 + x^2)$ in Fig. 3.4.9 .

3.   <u>Vertical asymptotes</u>. We are interested in the behavior on both sides of the x-values which produce a zero denominator. The sign of the limiting values on either side of a vertical asymptote may either be the same or opposite. Consider $f(x) = 1/x^2$ and $f(x) = 1/x$. Note that, unlike horizontal asymptotes, there may be any number of vertical asymptotes.

4.   <u>Other asymptotes</u>. Example 3 is interesting in that one of the asymptotes is neither vertical nor horizontal.

5.   <u>Six-step method</u>. There is no need to memorize the method. With enough practice, the method should become second nature to you.

6.   <u>Cusps</u>. These occur when the limit of the derivative becomes infinite, and when the sign of the derivative changes. Note the difference when the sign does not change, by comparing Figs. 3.4.15 and 3.4.16 . You need to be aware of the possibility of cusps only if fractional powers occur, such as $f(x) = x^{2/3}$.

SOLUTIONS TO EVERY OTHER ODD EXERCISE

1.   f is odd because $f(-x) = [(-x)^3 + 6(-x)]/[(-x)^2 + 1] = -(x^3 + 6x)/$

$(x^2 + 1) = -f(x)$.

5. $f(x) = x/(x^3 + 1)$ is undefined at $x = -1$, so it is a vertical asymptote.

For x near -1 and $x < -1$, x is negative and $x^3 + 1$ is small and

negative, so f(x) is large and positive. For x near -1 and $x > -1$,

x is negative and $x^3 + 1$ is small and positive, so f(x) is large and

negative.

9.

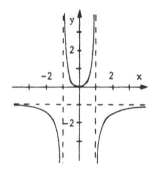

$f(-x) = f(x)$, so f is even and we
need to consider $x \geq 0$ only. $x = 1$ is
a vertical asymptote. For x near 1
and $x < 1$, f(x) is large and positive.
For x near 1 and $x > 1$, f(x) is
large and negative. $\lim\limits_{x \to \infty} f(x) =$
$\lim\limits_{x \to \infty} [1/(1/x^2 - 1)] = -1$, which is a
horizontal asymptote. $f'(x) = 2x/(1 - x^2)^2$,

so 0 is a critical point and f is increasing on (0,1) and (1,∞) .

$f''(x) = (2 + 6x^2)/(1 - x^2)^3$, so f'' has the same sign as $1 - x^2$.

There are no inflection points. $f''(0) = 2 > 0$, so 2 is a local minimum.

f is concave upward on (0,1) and concave downward on (1,∞) . Plot a few

points and use the symmetry to complete the graph.

13.

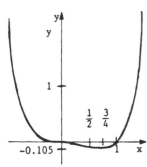

f(x) shows no symmetry. $f(x) = x^3(x - 1) = 0$ at $x = 0,1$. $f'(x) = x^2(4x - 3) = 0$ at $x = 0,3/4$. $f''(x) = 6x(2x - 1) = 0$ at $x = 0,1/2$; $f'''(x) = 6(4x - 1)$; $f'''(0) \neq 0$ and $f'''(1/2) \neq 0$, so 0 and 1/2 are inflection points. $f''(3/4) > 0$, so 3/4 is a local minimum. $f''(x) > 0$ for $x > 1/2$ and for $x < 0$; $f''(x) < 0$ for $0 < x < 1/2$; so f is concave upward on $(1/2,\infty)$ and $(-\infty,0)$, concave downward on $(0,1/2)$.

17.

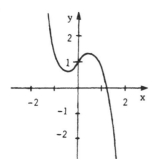

Even though f(x) is not symmetric about the y-axis or the origin, note that f(x) is the odd function $-x^3 + x$ shifted up 1 unit. $f'(x) = -3x^2 + 1 = 0$ at $x = \pm 1/\sqrt{3}$. $f''(x) = -6x = 0$ at $x = 0$. $f'''(x) = -6$, so $x = 0$ is an inflection point. $f''(-1/\sqrt{3}) > 0$, so $-1/\sqrt{3}$ is a local minimum and $(-\infty,0)$ is where f is concave up. $f''(1/\sqrt{3}) < 0$, so $1/\sqrt{3}$ is a local maximum and the graph is concave down on $(0,\infty)$.

21.

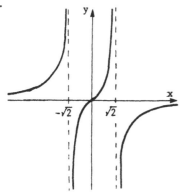

$f(x) = [x^3/(2 - x^2)] + x = [x^3 + x(2 - x^2)]/(2 - x^2) = 2x/(2 - x^2)$ is odd since $f(-x) = -f(x)$. Hence we only need to graph f for $x \geq 0$ (the graph for $x \leq 0$ can be obtained from that for $x \geq 0$ by reflecting the latter across the x-axis and then across the y-axis, or the y-axis first and then the x-axis). f has vertical asymptotes at

21 (continued).

x = ±√2 ; for x near √2 and x < √2 , f(x) is large and positive

for x near √2 and x > √2 , f(x) is large and negative. As |x|

gets large, $f(x) = (2x/x^2)/[(2 - x^2)/x^2] = (2/x)/[(2/x^2) - 1]$ approaches

0 , i.e., the line y = 0 is a horizontal asymptote. x = 0 is the

only zero of f . $f'(x) = 2(2 + x^2)/(2 - x^2)^2$ has no real zeros. Hence f

has no critical point. $f''(x) = [4x(x^2 + 6)]/(2 - x^2)^3 = [2(x^2 + 6)/$

$(2 - x^2)^2] f(x)$; so f″ has the same sign as f , i.e., f″ > 0 on

$(-\infty, -\sqrt{2})$ and $(0, \sqrt{2})$, f″ < 0 on $(-\sqrt{2}, 0)$ and $(\sqrt{2}, \infty)$, which implies

f is concave upward on $(-\infty, -\sqrt{2})$ and $(0, \sqrt{2})$, concave downward on

$(-\sqrt{2}, 0)$ and $(\sqrt{2}, \infty)$. $f'''(x) = 12(x^4 + 12x^2 + 4)/(2 - x^2)^4$. Hence

f″(0) = 0 but f‴(0) ≠ 0 , so x = 0 is an inflection point.

25.

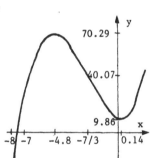

$f(x) = x^3 + 7x^2 - 2x + 10$ is neither odd
nor even, and is defined everywhere. $f'(x) =$
$3x^2 + 14x - 2 = 0$ at $x = (-7 ± \sqrt{55})/3 ≈$
0.14 , -4.8 , respectively. f′ is negative
between its roots and positive outside its
roots, so f is decreasing on (-4.8, 0.14) ,
increasing on $(-\infty, -4.8)$ and $(0.14, \infty)$. f′
changes sign (from positive to negative) at

-4.8 and (from negative to positive) at 0.14 , so x = -4.8 is a local

maximum with maximum value f(-4.8) ≈ 70.29 and x = 0.14 is a local

minimum with minimum value f(0.14) ≈ 9.86 . $f''(x) = 6x + 14 = 0$ at x =

-7/3 ; f‴(-7/3) = 6 ≠ 0 , so -7/3 is an inflection point.

29.

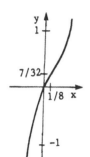

$f(x) = 8x^3 - 3x^2 + 2x = x(8x^2 - 3x + 2)$ is defined
everywhere, is neither odd nor even, and has only one
real zero (at $x = 0$) . $f'(x) = 2(12x^2 - 3x + 1)$
has no zero. $f''(x) = 48x - 6 = 0$ at $x = 1/8$,
$f'''(1/8) = 48 \neq 0$, so $1/8$ is an inflection point
with $f(1/8) = 7/32$. $f''(x) > 0$ for $x > 1/8$ and
$f''(x) < 0$ for $x < 1/8$. Hence f is concave upward on $(1/8, \infty)$ and concave downward on $(-\infty, 1/8)$.

33. (a) $f'(x) = (2/3)(x^2 - 3)^{-1/3}(2x) = 4x/3(x^2 - 3)^{1/3}$. $f'(x) = 0$ only
at $x = 0$, and $f'(x)$ is undefined at $x = \pm\sqrt{3}$. $f'(x) > 0$ on
$(-\sqrt{3}, 0)$ and $(\sqrt{3}, \infty)$, the intervals where f is increasing. On
$(-\infty, -\sqrt{3})$ and $(0, \sqrt{3})$, $f'(x) < 0$ and f is decreasing.

(b)

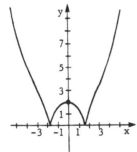

$f(x)$ is an even function. This is a
nonnegative function with zeros at $\pm\sqrt{3}$.
$f''(x) = 4(x^2 - 9)/9(x^2 - 3)^{4/3}$, so inflection points occur at $x = \pm 3$. Cusps
occur at $x = \pm\sqrt{3}$.

37.

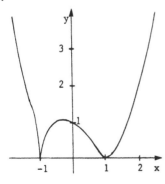

$f(x) = (x - 1)^{4/3}(x + 1)^{2/3}$; $f'(x) =$
$(4/3)(x - 1)^{1/3}(x + 1)^{2/3} + (2/3)(x - 1)^{4/3} \times$
$(x + 1)^{-1/3} = (x - 1)^{1/3}(x + 1)^{-1/3}(4x + 4 +$
$2x - 2)/3 = (2/3)(x - 1)^{1/3}(x + 1)^{-1/3}(3x + 1)$
$f'(x) = 0$ if $x = 1, -1/3$. $\lim_{x \to -1}[1/f'(x)] =$
0 , so a cusp exists at $x = -1$. $f''(x) =$
$(2/3)(x + 1)^{-1/3}[(1/3)(x - 1)^{-2/3}(3x - 1) +$
$(x - 1)^{1/3} \cdot 3] + (2/3)(-1/3)(x + 1)^{-4/3}(x - 1)^{1/}$

37 (continued).

$(3x - 1) = (2/9)(x + 1)^{-4/3}(x - 1)^{-2/3}(15x^2 + 10x - 13)$. $f''(x) = 0$ if

$15x^2 + 10x - 13 = 0$,   i.e., $x = (-10 \pm \sqrt{880})/30 \approx 0.655, -1.322$; $f''(x)$

is not defined at $x = \pm 1$.   f   is increasing on $(-1, -1/3), (1, \infty)$,   and

decreasing on $(-\infty, -1), (-1/3, 1)$.   The graph is concave upward on

$(-\infty, -1.322), (0.655, \infty)$,   and concave downward on $(-1.322, 0.655)$.

41.   f   is even if and only if $f(x) = f(-x)$,   i.e., $f(x) - f(-x) = 0$

for all   x .   For $f(x) = a_n x^n + a_{n-1} x^{n-1} + \ldots + a_1 x + a_0$, $f(-x) =$

$a_n(-x)^n + a_{n-1}(-x)^{n-1} + \ldots + a_1(-x) + a_0$.   Thus, $f(x) - f(-x) = 2a_1 x +$

$2a_3 x^3 + \ldots = 0$ for all x if and only if $a_1 = a_3 = a_5 = \ldots = 0$,

i.e.,   f   is even if and only if the even powers of   x   occur with

nonzero coefficients in $f(x)$.   The case where $f(x)$ is odd is done

in a similar fashion.

45. $f(x) = e(x) + o(x)$, $f(-x) = e(-x) + o(-x) = e(x) - o(x)$ since   e   is

even and   o   is odd.   Adding and subtracting these two equations yield $f(x) +$

$f(-x) = 2e(x)$, $f(x) - f(-x) = 2o(x)$;   so $e(x) = [f(x) + f(-x)]/2$

and $o(x) = [f(x) - f(-x)]/2$.   We check that $[f(x) + f(-x)]/2$ and

$[f(x) - f(-x)]/2$ are indeed even and odd, respectively, and that their

sum is $f(x)$.

49. $f(x) = ax^3 + bx^2 + cx + d$ $(a \neq 0)$; $f'(x) = 3ax^2 + 2bx + c$; $f''(x) =$

$6ax + 2b = 0$ at $x = -b/3a$; $f'''(-b/3a) = 6a \neq 0$,   so   f   has an

inflection point at $x = -b/3a$ and nowhere else. $g(x) = f(x - b/3a) -$

$f(-b/3a) = a(x - b/3a)^3 + b(x - b/3a)^2 + c(x - b/3a) + d - a(-b/3a)^3 -$

$b(-b/3a)^2 - c(-b/3a) - d = ax^3 + (c - b^2/3a)x$ is odd and $g(0) = 0$

(two properties which   g   would not have possessed had   g   been not

so defined as to eliminate the quadratic and the constant terms).

49 (continued).

$g'(x) = 3ax^2 + (c - b^2/3a)$; $g''(x) = 6ax = 0$ only at $x = 0$. $g'''(0) =$

$6a \neq 0$, so g has an inflection point at $x = 0$ and nowhere else.

Hence the point $(0, g(0)) = (0,0)$ acts for g as $(-b/3a, f(-b/3a))$

did for f . As f and g are cubics with the same leading coeffi-

cients, their graphs have the same basic shape. Hence f is symmetric

about its inflection point $(-b/3a, f(-b/3a))$ provided g is symmetric

about its inflection point $(0, g(0)) = (0,0)$ which is the case since

g is odd.

53.

The velocity is given by $x'(t) = (5/3) \times$
$(t - 1)^{2/3}$ so $\lim_{t \to 1} [x'(t)] = 0$. The
acceleration is $x''(t) = (10/9)(t - 1)^{-1/3}$,
so $\lim_{t \to 1} [x''(t)]$ is infinite. See the graphs
of $x'(t)$ and $x''(t)$ at the left. The graph
of $x(t)$ is shown in the answer section of
the text.

57. $f(x - b/4a) = a(x - b/4a)^4 + b(x - b/4a)^3 + c(x - b/4a)^2 + d(x - b/4a) +$

$e = a(x^4 - bx^3/a + 3b^2x^2/4a^2 - b^3/16a^3 + b^4/256a^4) + b(x^3 - bx^2/4a +$

$bx/16a^2 - b^3/64a^3) + c(x^2 - bx/2a + b^2/16a^2) + d(x - b/4a) + e = ax^4 +$

$(-b + b)x^3 + (3b^2/4a - b^2/4a + c)x^2 + (-b^2/16a^2 + b^2/16a^2 + cb/2a + d)x +$

$(b^4/256a^3 - b^4/64a^3 + cb^2/16a^2 + db/4a + e)$. Dividing by a and using

new coefficients, we have $f(x) = x^4 + c_1 x^2 + d_1 x + e$.

61. A type I quartic occurs in the right half-plane where $c > 0$. A type II

quartic occurs on the d-axis. Type II_1 occurs on the positive d-axis; type I

occurs at the origin, and type II_3 occurs on the negative d-axis. A

type III quartic occurs in the left half-plane where $c < 0$. Type III_1

occurs in the upper white region where $d > -(4c/3)\sqrt{-c/6}$; type III_2

occurs on the curve which forms the upper boundary of the gray region

61 (continued).

where $d = -(4c/3)\sqrt{-c/6}$;   type III_3 occurs in the gray region where $|d| < |(4c/3)\sqrt{-c/6}|$;   type III_4 occurs on the curve which forms the lower boundary of the gray region where $d = (4c/3)\sqrt{-c/6}$,   and type III_5 occurs in the lower white region where $d < (4c/3)\sqrt{-c/6}$.

SECTION QUIZ

1. Consider the function $f(x) = x^3/(1 - x^2)$

 (a) Discuss the symmetries of this function.

 (b) Where are the asymptotes of this function?

 (c) Sketch the graph.

2. Sketch the graph of $f(x) = (1 - x^{2/3})^3$. Clearly label the critical points, the inflection points, and the cusps.

3. A brilliant piratess had left a treasure map for her husband. Clues describing the exact location of a golden treasure were left along the curves described by $y = (x^3 - 1)/(x^3 - 3)$. The origin is located at the center of their island home and each unit represents 10 paces. Make a sketch of the paths where more clues may be found. Discuss the symmetry, asymptotes, increasing-decreasing, and concavity.

SOLUTIONS TO PREREQUISITE QUIZ

1. $f'(x_0) = 0$

2. $g'(x_0) > 0$

3. Increasing

4. Concave downward

5.   (a)   2

     (b)   3

     (c) $+\infty$

     (d) $-\infty$

SOLUTIONS TO SECTION QUIZ

1.   (a)   Symmetric about origin

     (b) $x = \pm 1$ (Also, $y = -x$)

     (c)

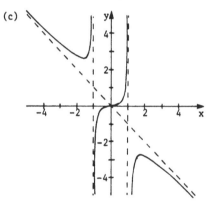

2.

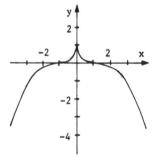

Critical and inflection points at $(\pm 1, 0)$;

cusp at $(0, 1)$.

3.

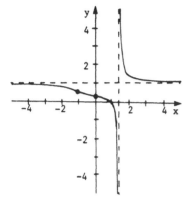

There are no special symmetries;
$x = \sqrt[3]{3}$ is a vertical asymptote;
$y = 1$ is a horizontal asymptote;
decreasing for $x \neq \sqrt[3]{3}$; concave
upward on $(-\sqrt[3]{3/2},0),(\sqrt[3]{3},\infty)$;
concave downward on $(-\infty,-\sqrt[3]{3/2})$,
$(0,\sqrt[3]{3})$. There are no critical
points; however, inflection points
occur at $x = -\sqrt[3]{3/2}$ and at $x = 0$.

3.5   Maximum-minimum Problems

PREREQUISITES

1.   Recall how to find critical points (Section 3.2).

2.   Recall tests used to determine maxima and minima (Section 3.2).

PREREQUISITE QUIZ

1.   Classify the critical points of the following as local minima, local

maxima, or neither:

(a) $(x^2 - 3)/(x - 2)$

(b) $(x + 3)(x^2 - x - 1)$

2.   If $f(x_0)$ is a local minimum, what can you say about $f'(x_0)$? What

about $f''(x_0)$?

GOALS

1.   Be able to find the global minimum and the global maximum on an interval.

2.   Be able to solve minimum-maximum word problems.

STUDY HINTS

1.   Definitions. A local maximum is a maximum on a small interval, whereas

a global maximum is a maximum on the entire interval of definition.

Note that a global maximum is always a local maximum, but not vice

versa. Similar comments may be made for minima.

2.   Points and values. The value of a maximum refers to   y   and points

refer to   x . Whereas there can only be one value for the maximum,

there may be several points where the maximum is assumed.

3. <u>Extreme value theorem</u>. Note that two conditions are necessary:

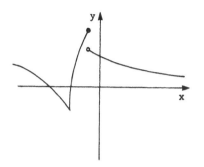

(i) continuity and (ii) closed intervals. By eliminating just one condition, the guarantee of a minimum and
maximum no longer exists. However,
both minimum and maximum may occur
even if one or none of the conditions
hold. See figure.

4. <u>Closed interval test</u>. Essentially, it is saying, "If a function is
(i) continuous and (ii) differentiable on a closed interval, then the
minimum and maximum points must exist at-critical points or endpoints."
Think about what happens if differentiability wasn't required, as in
$f(x) = |x|$. What if a discontinuity existed? What if the interval
was not closed?

5. <u>Solving word problems</u>. The most difficult part is translating words
into equations. Whenever possible, drawing a picture often helps one
to visualize and understand the entire question. Write down equations
to describe the situation. If you have several variables, try to re-
write them in terms of a single variable. Before finding a solution,
a guess may help determine the correctness of your answer. After
finding a solution, ask yourself if it makes sense. Finally, we
emphasize that practice makes perfect.

6. <u>Test taking</u>. Calculus exams almost always have a maximum-minimum word
problem. In the majority of cases, you will be asked to solve problems
with a minimum of information. Thus, you should try to use all of the

6.   Test taking (continued).

numbers and equations which are given to you. You should also be aware

that some instructors attempt to test your understanding by giving

extraneous information.

SOLUTIONS TO EVERY OTHER ODD EXERCISE

1.   Use a variation of the closed inteval test and analyze $\lim\limits_{x \to \infty} f(x)$ along

with the critical and endpoints. $f'(x) = 6x^2 - 5$, so the critical

points are $x = \pm\sqrt{5/6}$; however, neither critical point is in $[1, \infty)$.

$f(1) = -1$ and $\lim\limits_{x \to \infty} f(x) = \infty$. Thus, 1 is the minimum point and -1

is the minimum value. There is no maximum point.

5.   Use a variation of the closed interval test and analyze $\lim\limits_{x \to -\infty} f(x)$,

$\lim\limits_{x \to 0} f(x)$, and the critical points. $f(x) = x + 1/x$, so $f'(x) =$

$1 - 1/x^2$, so the critical points are ± 1 . $\lim\limits_{x \to -\infty} f(x) = -\infty$, $f(-1) =$

-2, and when x is near 0 and negative, the limit is $-\infty$. Thus,

the maximum point is -2 and there is no minimum point in $(-\infty, 0)$.

9.   The minimum point is about 1971.4 and the maximum point is 1980 .

The minimum value is about 16% , whereas the maximum value is about 34% .

13. $f'(x) = 4x^3 - 8x$, so $x = 0, \pm\sqrt{2}$ are the critical points; -4 and 2

are the endpoints. $f(-4) = 199$; $f(-\sqrt{2}) = 3$; $f(0) = 7$; $f(\sqrt{2}) = 3$;

$f(2) = 7$. Therefore, -4 is the maximum point; the maximum value is

199 . $\pm\sqrt{2}$ are the minimum points and 3 is the minimum value.

17. Use a variation of the closed interval test and analyze the limits at

$\pm\infty$, if necessary. $f'(x) = 2x - 3$, so the critical point is 3/2 .

$\lim\limits_{x \to -\infty} f(x) = \infty$; $f(-8) = 89$; $f(-3) = 19$; $f(-2) = 11$; $f(-3/2) = 7.75$;

$f(-1) = 5$; $f(1/2) = -0.25$; $f(1) = -1$; $f(3/2) = -1.25$; $f(2) = -1$;

$f(8) = 41$; $\lim\limits_{x \to \infty} f(x) = \infty$.

17 (continued).

(a)   There are no critical points, endpoints, nor minimum and maximum points.

(b)   There is no critical point. The endpoint is $1/2$, which is the minimum point. The minimum value is $-1/4$. There is no maximum point.

(c)   The critical point is $3/2$ and the endpoint is 2 . The minimum point is $3/2$ with minimum value $-5/4$. There is no maximum point.

(d)   The answer is the same as part (c).

(e)   The critical point is $3/2$. There are no endpoints. The minimum point is $3/2$ with minimum value $-5/4$. There is no maximum point.

(f)   The answer is the same as part (e) .

(g)   There is no critical point. The endpoints are -1 and 1 . The maximum point is -1 and the minimum point is 1 . The maximum value is 5 and the minimum value is -1 .

(h)   The critical point is $3/2$. The endpoints are -8 and 8 . The maximum point is -8 and the minimum point is $3/2$. The maximum value is 89 and the minimum is $-5/4$.

21. $f'(x) = 3x^2 - 6x + 3 = 3(x^2 - 2x + 1) = 3(x - 1)^2$. $f'(x) = 0$ at $x = 1$, so this is the critical point. There are no endpoints. Since $\lim_{x \to -\infty} f(x) = -\infty$ and $\lim_{x \to +\infty} f(x) = +\infty$, there are no maximum or minimum points, nor any maximum or minimum values.

25. $f'(x) = [3x^2(x^2 + 1) - (x^3 - 1)(2x)]/(x^2 + 1)^2 = (x^4 + 3x^2 + 2x)/$

   $(x^2 + 1)^2 = x(x^3 + 3x + 2)/(x^2 + 1)^2$. Using a calculator, we find

   that the critical points are 0 and -0.60 . The endpoints are -10

   and 10 . $f(-10) = -1001/101$; $f(-0.60) \approx -0.89$; $f(0) = -1$;

   $f(10) = 999/101$, so the maximum point is 10 and the maximum value

   is 999/101 . The minimum point is -10 and the minimum value is

   $-1001/101$.

29. $f'(x) = [3x^2(1 + x^2) - x^3(2x)]/(1 + x^2)^2 = (x^4 + 3x^2)/(1 + x^2)^2$,

   which equals 0 at $x = 0$, so this is the critical point. There

   are no endpoints. $\lim\limits_{x \to -\infty} f(x) = -\infty$ and $\lim\limits_{x \to +\infty} f(x) = +\infty$, so there are

   no maximum or minimum points.

33. Let the two individual masses be x and y . We are given that

   $x + y = M$, and that the force is proportional to xy . We want to

   maximize xy , i.e., maximize $f(x) = x(M - x)$.

37. $f(x) = x(M - x)$, so $f'(x) = M - 2x = 0$ if $x = M/2$. Also, $f''(M/2) =$

   $-2 < 0$, which implies that $x = M/2$ is a maximum point. Hence, the

   two masses should be equal.

41. (a)   We are given $V = \pi r^2 h = 1000$ and we want to

   minimize the surface area, $A = 2(\pi r^2) + 2\pi rh$.

   $\pi r^2 h = 1000$ implies $h = 1000/\pi r^2$, so $A =$

   $2\pi r^2 + 2000/r$. Thus, $A'(r) = 4\pi r - 2000/r^2 =$

   0 implies $4\pi r^3 = 2000$, so $r = \sqrt[3]{500/\pi}$ cm. and $h = 1000/$

   $\pi(500/\pi)^{2/3} = 2\sqrt[3]{500/\pi}$ cm.

   (b) $V = \pi r^2 h$ implies $h = V/\pi r^2$, so $A = 2\pi r^2 + 2V/r$. Therefore,

   $A'(r) = 4\pi r - 2V/r^2 = 0$ implies $4\pi r^3 = 2V$, so $r = \sqrt[3]{V/2\pi}$ cm.

   and $h = V/\pi(V/2\pi)^{2/3} = \sqrt[3]{4V/\pi} = 2\sqrt[3]{V/2\pi}$ cm.

41 (continued).

(c)   Here,   A   is constant and we want to maximize   V . $A = 2\pi r^2 +$
$2\pi rh$ implies $h = (A - 2\pi r^2)/2\pi r$,   so $V = (A - 2\pi r^2)r/2 =$
$(Ar - 2\pi r^3)/2$ and $V'(r) = (A - 6\pi r^2)/2$.   Thus, $V'(r) = 0$
when $r = \sqrt{A/6\pi}$ cm. and $h = (2A/3)/(2\pi\sqrt{A/6\pi}) = A\sqrt{6\pi}/3\pi\sqrt{A} =$
$\sqrt{2A/3\pi}$ cm.

45.   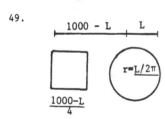   We are given that $\ell + 2h + 2h = \ell + 4h = 72$,
and we want to maximize $V = \ell h^2$. $\ell + 4h = 72$
implies $\ell = 72 - 4h$,   so $V = (72 - 4h)h^2 =$
$72h^2 - 4h^3$,   and $V'(h) = 144h - 12h^2 = 12h(12 - h)$.

Thus, $V'(h) = 0$ when $h = 0$ or $h = 12$,   but $h = 0$ means $V = 0$,
so h must be   12 .   Thus, the height and width are   12   inches, while
the length is   24   inches.

49.   1000 – L   L   Use $(1000 - L)$ feet of fencing for the
square area and   L   feet of fencing for the
$r = L/2\pi$ circular area.   Each side of the square is
250 – L/4   and the circle has a radius of
$\frac{1000-L}{4}$ $L/2\pi$.   Therefore, the total area, $A =$
$(250 - L/4)^2 + \pi(L/2\pi)^2$,   so $A'(L) =$
$2(250 - L/4)(-1/4) + \pi(2)(L/2\pi)(1/2\pi) =$

$-125 + L/8 + L/2\pi = (-1000\pi + L\pi + 4L)/8\pi$. $A'(L) = 0$ when $L =$
$1000\pi/(\pi + 4)$. $A''(L) = 1/8 + 1/2\pi > 0$,   so the minimum area is
attained at the critical point.

(a)   The only critical point is a minimum, so the maximum occurs at the
endpoints.   When all of the fencing is used for the square area,
the total area is $(250)^2 = 62{,}500$ square feet.   If all of the
fencing is used for the circular area, the total area is

49 (continued).

   (a)(continued) $\pi(500/\pi)^2 = 250{,}000/\pi \approx 79{,}577.5$ square feet, so

      the maximum occurs when all of the fencing is used for the circular

      area, which has a radius of $(500/\pi)$ feet.

   (b)   The minimum area is attained at the critical point. The side of

      the square is $[250 - 250\pi/(\pi + 4)] = [1000/(\pi + 4)]$ feet and

      the radius of the circle is $[500/(\pi + 4)]$ feet. The total area

      is $[1000/(\pi + 4)]^2 + \pi[500/(\pi + 4)]^2 = [4(500)^2 + \pi(500)^2]/$

      $(\pi + 4)^2 = [250{,}000/(\pi + 4)]$ square feet.

53.

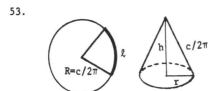

From geometry, we know that $c = 2\pi R$, so $R = c/2\pi$, which is the slant height of the dunce cap. The circumference of the cap's base is $c - \ell$, so $r = (c - \ell)/2\pi$, and we also know

that $h^2 = (c/2\pi)^2 - r^2$. We want to maximize $V = \pi r^2 h/3$, but it is

easier to maximize $V^2 = \pi^2 r^4 h^2/9 = \pi^2[(c - \ell)/2\pi]^4[(c/2\pi)^2 -$

$((c - \ell)/2\pi)^2]/9 = (c - \ell)^4(2c\ell - \ell^2)/576\pi^4$. Thus, $(V^2)'(\ell) =$

$[4(c - \ell)^3(-1)(2c\ell - \ell^2) + (c - \ell)^4(2c - 2\ell)]/576\pi^4 = 2(c - \ell)^3(-4c\ell +$

$2\ell^2 + c^2 - 2c\ell + \ell^2)/576\pi^4 = 2(c - \ell)^3(3\ell^2 - 6c\ell + c^2)/576\pi^4$. To

maximize V^2, we solve $0 = 3\ell^2 - 6c\ell + c^2$, which has the solutions

$\ell = (6c \pm 2\sqrt{6}c)/6$. $(3 + \sqrt{6})c/3$ gives us $r < 0$, so the answer is

$\ell = (3 - \sqrt{6})c/3$.

57. This problem requires the maximizations of area. If we reflect the

    shape of the fencing across the shoreline, we have a symmetric figure.

    A circle, the most symmetric of all geometric figures, should maximize

    area if the perimeter is held constant. Therefore, the 500 feet of

57 (continued).

fencing should be used to make a semicircle with a radius of $(500/\pi)$

meters.   The maximum area is $(125,000/\pi)$ square meters.

61.

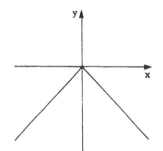

One possibility is a function which comes
to a point at its maximum.   A simple
example is $y = -|x|$.   There is no
unique solution.

65.   Suppose M_1 and M_2 are both maximum values of f on I .   So there

exists at least one x_1 in I such that $f(x_1) = M_1$. M_2 being a

maximum value implies $M_2 \geqslant f(x)$ for all x in I ,   in particular

$M_2 \geqslant f(x_1) = M_1 (Eq.1)$.

On the other hand, M_2 being a maximum value of f on I implies

that there exists an x_2 in I such that $f(x_2) = M_2$. M_1 being a

maximum value implies $M_1 \geqslant f(x_2) = M_2 (Eq.2)$.   Combining (Eq. 1) and (Eq. 2)

we have $M_1 = M_2$,   i.e., the maximum value of f in I is unique.

69. $f'(x) = [(1 - x^2) - x(-2x)]/(1 - x^2)^2 = (1 + x^2)/(1 - x^2)^2$ for

$-1 < x < 0$ and $f'(x) = 3x^2 - 1$ for $0 < x < 2$. $f'(0)$ does not

exist.   The critical point is $\sqrt{1/3}$ and there are no endpoints.   For

x near -1 and $x > -1$, $\lim\limits_{x \to -1} f(x) = -\infty$; $f(0) = 0$; $f(\sqrt{1/3}) =$

$-2/3\sqrt{3} \approx -0.38$; $\lim\limits_{x \to 2} f(x) = 6$.   Since $f(x)$ only approaches   6 ,

$f(x)$ has no maximum nor minimum values.

73. (a)   At the supermarket, the cheapest price is 32.5 cents per quart if
    one buys a half gallon. At the corner grocery, a half gallon of
    milk costs 80 cents; however, one can buy a quart today and then
    a half gallon tomorrow which costs 77.5 cents for the first two
    quarts, so a quart should be purchased today.

    (b)   As shown in part (a), a half gallon should be bought if $g/2 < q$
    and $G < Q + g/2$. If $q < g/2$, as when a sale occurs, then a
    half gallon should be bought if $G < Q + q$.

77. By the definition of concavity (Section 3.3), f is differentiable on
    (a,b) and f lies above its tangent lines, locally. Thus, for x
    near x_0 , we have $f(x) > f(x_0) + f'(x_0)(x - x_0)$. Since f is
    differentiable, the closed interval test tells us that the maximum
    must be a critical point or an endpoint. If x_0 is a critical point,
    then $f(x) > f(x_0)$, so x_0 is a minimum. Therefore, the maximum
    point of f is an endpoint.

SECTION QUIZ

1.   This question demonstrates the limitations of the extreme value theorem.

    (a)   Sketch the graph of a function on a closed interval which lacks a
    minimum and a maximum.

    (b)   Sketch the graph of a continuous function which has neither a
    minimum nor a maximum.

    (c)   Sketch the graph of a discontinuous function on an open interval
    which contains both a minimum and a maximum.

2.   True or false:

    (a)   A global maximum must be a local maximum.

    (b)   If a function contains several local minima, the smallest of these
    is the global minima.

3.   Crazy Charlie challenged a city bus to a demolition derby.   Crazy

Charlie's little compact will have a value of $c(t) = 12000 - 30t^2$,

t   minutes into the derby.   The value of the bus will be $b(t) =$

$20000 - 10t^2 - 80t$.   Crazy Charlie quits when his car is totaled and

has a value of zero.   Find the minimum and maximum differences between

the values of the vehicles, $|c(t) - b(t)|$.

SOLUTIONS TO PREREQUISITE QUIZ

1.   (a)   local maximum: 1 ;   local minimum: 3

    (b)   local maximum: -2 ;   local minimum: 2/3

2. $f'(x_0) = 0$; $f''(x_0) \geqslant 0$.

SOLUTIONS TO SECTION QUIZ

1.   (a)

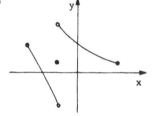

    (b)

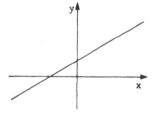

    (c)

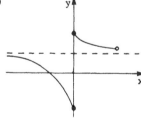

2.   True.

3.   Maximum = \$14400 ;   minimum = \$7920 .

3.6   The Mean Value Theorem

PREREQUISITES

1.   Recall how to find an average rate of change (Section 1.1).

2.   Recall the concept of continuity (Sections 1.2 and 3.1).

3.   Recall the critical point test (Section 3.2).

PREREQUISITE QUIZ

1.   A tour bus travels $4t^{3/2}$ kilometers in t hours. What is the average velocity during the interval $1 \leqslant t \leqslant 4$?

2.   If $f(x_0)$ exists, what other condition is necessary for continuity?

3.   If $f(x_0)$ is a local minimum or a local maximum, what can you say about $f'(x_0)$, assuming that $f'(x)$ exists?

GOALS

1.   Be able to understand the mean value theorem.

2.   Be able to estimate the difference between function values, given information about the derivative.

STUDY HINTS

1.   Mean value theorem. The theorem states that the average slope is attained somewhere in an interval, possibly at more than one point, if two conditions hold: i) the function is continuous on $[a,b]$ and ii) it is differentiable on (a,b). Example 3 explains what happens if differentiability doesn't hold.

2.   **Typical application.** Example 2 shows a typical proof which relies on
     getting an expression of the form $[f(x_1) - f(x_0)]/(x_1 - x_0)$. Many
     times, as in Example 2, you will simply be asked to multiply through
     by the denominator to estimate the difference between function values.

3.   **Rolle's theorem.** Basically, it asserts that if $f = 0$ at two points,
     then the derivative is zero somewhere in between, assuming continuity
     and differentiability.

4.   **Horserace theorem.** This says that if two continuous, differentiable
     functions have equal values at the endpoints, then they must have equal
     slopes somewhere in between.

SOLUTIONS TO EVERY OTHER ODD EXERCISE

1.   Since f is continous on $[0,1/2]$ and $f'(x)$ exists for x in
     $(0,1/2)$, the mean value theorem applies. It states that there exists
     x_0 in $(0,1/2)$ such that $f'(x_0) = [f(1/2) - f(0)]/(1/2 - 0)$. Since
     $0.3 \leq f'(x) \leq 1$ for x in $(0,1/2)$, we have, in particular,
     $0.3 \leq f'(x_0) \leq 1$, i.e., $0.3 \leq [f(1/2) - f(0)]/(1/2) \leq 1$ or $0.15 \leq$
     $f(1/2) - f(0) \leq 1/2$.

5.   Since $f(x)$ is a polynomial, it is continous and differentiable. By
     the mean value theorem, there is an x_0 in $(-1,0)$ such that $f'(x_0) =$
     $[f(0) - f(-1)]/(0 - (-1)) = (15 - 17)/1 = -2$.

9.   Let $x(t)$ be the position of the object. We are given that the velocity
     of the object $x'(t)$ is such that $34 \leq x'(t) \leq 36$. Antidifferentiation
     gives $34t + C \leq x(t) \leq 36t + C$. $x(0) = 4$ gives $C = 4$. Hence,
     $34t + 4 \leq x(t) \leq 36t + 4$ and so $72 \leq x(2) \leq 76$.

13. The derivative of $1/x$ is $-(1/x^2)$, so by Consequence 3, $F(x) = (1/x) + C$, where C is a constant, on any interval not containing zero. So $F(x) = (1/x) + C_1$ for $x < 0$ and $F(x) = (1/x) + C_2$ for $x > 0$. Hence C_1 and C_2 are constants, but they are not necessarily equal.

17. By the power and polynomial rules for antidifferentiation, we get
$(1/2)x^2/2 - 4x^3/3 + 21x + C = x^2/4 - 4x^3/3 + 21x + C$.

21. By the power rule for antidifferentiation, we get $F(x) = 2x^5/5 + C$, but $F(1) = 2/5 + C = 2$ implies $C = 8/5$. Thus, $F(x) = 2x^5/5 + 8/5$.

25. Let c, d, and e be the points with $a < c < d < e < b$ in (a,b) at which f vanishes, then $[f(d) - f(c)]/(d - c) = 0$ and $[f(e) - f(d)]/(e - d) = 0$ implies that the first derivative vanishes somewhere in (c,d) and somewhere in (d,e) according to the mean value theorem. Now, let m and n be the points in (a,b) where $f'(x)$ vanishes. By the mean value theorem, there exists an x_0 in (m,n) such that $f''(x_0) = [f'(n) - f'(m)]/(n - m) = 0$. Since (m,n) is in (a,b), the statement is proven.

29. By applying the mean value theorem to $[t_1,t_2]$, we have $[N(t_2) - N(t_1)]/(t_2 - t_1) = 0$ because $N(t_2) = N(t_1)$. Therefore, dN/dt cannot be greater than 0 on all of (t_1,t_2) . Also, N is not constant, so dN/dt cannot be 0 on the entire interval. Consequently, $dN/dt < 0$ for some T in (t_1,t_2) , which means that the population is decreasing. Do the same for the interval $[t_2,t_3]$.

SECTION QUIZ

1. (a) What are the two important conditions which are required before
         the mean value theorem can be applied?

    (b) What equation describes the statement of the mean value thoerem?

    (c) Suppose $f(x) = (x^3 - 3)/(x + 2)$. Show that somewhere in $(0,2)$,
         the slope of $f(x)$ is $11/8$.

2. Consider the function $g(t) = 1/t^2$ on $(-1,3)$. Then $[g(3) - g(-1)]/$
    $(3 - (-1)) = -2/9$. Now, $g'(t) = -2/t^3$, which is $-2/9$ at $t_0 =$
    $\sqrt[3]{9}$. How can the mean value theorem be satisfied when there is a dis-
    continuity at $t = 0$?

3. Mean Michael's favorite pastime was playing with little girl's ponytails.
    When he was in the third grade, Michael would sneak up behind his school-
    mates during recess, yank their ponytails, and tie their hair to a low-
    hanging tree branch. Michael's technique imporved as the week went by
    and during the weekend, he became well-behaved with his parents. After t
    days of school during the week, Mean Michael plays with approximately
    t^2 girl's ponytails.

    (a) Use the mean value theorem to show that at some time during the
         week, he yanks ponytails at a rate of about 5 per day.

    (b) Approximately when is the rate about 5 per day?

ANSWERS TO PREREQUISITE QUIZ

1. $28/3$ km/hr .

2. $\lim\limits_{x \to x_0} f(x) = f(x_0)$.

3. $f'(x_0) = 0$

ANSWERS TO SECTION QUIZ

1.   (a)   Continuity and differentiability .

     (b) $f'(x_0) = [f(b) - f(a)]/(b - a)$.

     (c)   Use the mean value theorem. $f(x)$ is continuous and differentiable

           on $(0,2)$; $[f(2) - f(0)]/2 = 11/8$.

2.   This is a case in which the conclusion is true even though the hypotheses

     are not met. Meeting the hypotheses guarantees the conclusion, but a

     conclusion may be true even without meeting the hypotheses.

3.   (a) $[f(5) - f(0)]/(5 - 0) = 5$.

     (b)   At about $5/2$ days into the week.

3.R   Review Exercises for Chapter 3

SOLUTIONS TO EVERY OTHER ODD EXERCISE

1. The function is continuous whenever the denominator exists and does not
    equal zero. We need $x^2 - 1 > 0$, so $f(x)$ is continuous in $(-\infty,-1)$
    and $(1,\infty)$.

5. $h(2) = -(2)^3 + 5 = -3$. From the left, $\lim\limits_{x \to 2-} h(x) = -(2)^3 + 5 = -3$.
    From the right, $\lim\limits_{x \to 2+} h(x) = (2)^2 - 7 = -3$. The limit of h at $x = 2$
    exists and is equal to the value of h at $x = 2$, so h is con-
    tinuous at $x = 2$. h is continuous on the whole real line because
    the function is continuous at $x = 2$ and all polynomials are continuous.

9. $f(0) = -2$ and $f(1) = 1$. By the intermediate value theorem (version 1),
    $f(x)$ has a root in $(0,1)$.

13. $f'(x) = 24x^2 - 6x = 6x(4x - 1) = 0$ at $x = 0,1/4$. $f'(x) > 0$ on $(-\infty,0)$
    and $(1/4,\infty)$, so it is increasing on these intervals. $f'(x) < 0$ on
    $(0,1/4)$, so it is decreasing on this interval.

17. $B(t) = t^3/10 - t^2/2 - 5t/2 + 10$, $-3 \leqslant t \leqslant 7$ (in this particular prob-
    lem, a negative value of t is possible for it is used to represent a
    time before the reference time, January 1, 1993). $B'(t) = 3t^2/10 - t -$
    $5/2 = 0$ at $t = -5/3$ and 5 in $[-3,7]$. $B'(t) < 0$ on $(-5/3,5)$;
    $B'(t) > 0$ on $[-3,-5/3)$ and $(5,7]$. So B is decreasing on $(-5/3,5)$,
    increasing on $[-3,-5/3)$ and $(5,7]$. $B''(t) = 3t/5 - 1 = 0$ at $t = 5/3$;
    $B'''(5/3) = 3/5 \neq 0$, so $5/3$ is an inflection point. $B'''(t) > 0$ for
    t in $(5/3,7]$ and $B''(t) < 0$ for t in $[-3,5/3)$. So 5 is a local
    minimum and $-5/3$ is a local maximum. 1990 will begin with a high infla-
    tion rate of 10.3% that will keep on increasing until it hits a record
    high of 15.1% (maximum) approximately one and one-third years later
    when luck begins to turn, i.e., the inflation rate will begin to de-
    crease. To the joys of the politicians, this trend will continue steadily,

17 (continued).

with even an accelerated decrease in September of 1994 (inflection point),
and reaching their goal of 0% inflation rate in 1996. So they will be
quite surprised to see a change of events around 1995 when the inflation
rate begins to increase for the first time since mid-1991. This increase
continues to the year 2000.

21. Since $f'(x) = 6x^2 - 10x + 4 = 2(3x - 2)(x - 1)$, the critical points
are 2/3 and 1 . $f''(x) = 12x - 10$. Since $f''(2/3) = -2 < 0$, 2/3
is a local maximum. And since $f''(1) = 2 > 0$, 1 is a local minimum.

25.

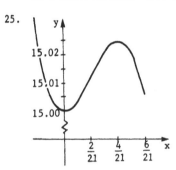

f(x) is a polynomial, so it is continuous
on $(-\infty,\infty)$, and it is differentiable on
 $(-\infty,\infty)$. $f'(x) = -21x^2 + 4x = x(-21x + 4)$,
so the critical points are x = 0,4/21 .
 $f''(x) = -42x + 4 = 0$ at x = 2/21 .
 $f'''(x) = -42$. Thus f is increasing on
 $(0,4/21)$ and decreasing on $(-\infty,0)$ and

 $(4/21,\infty)$. It is concave upward on $(-\infty,2/21)$ and concave downward on
 $(2/21,\infty)$. There are no endpoints. There is a local maximum at 4/21
and a local minimum at 0 . 2/21 is an inflection point.

29.

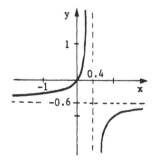

The domain is $x \neq 2/5$. f(x) is continuous and differentiable on $(-\infty,2/5)$ and
 $(2/5,\infty)$. $f'(x) = [3(2 - 5x) - 3x(-5)]/$
 $(2 - 5x)^2 = 6/(2 - 5x)^2$, so there are no
critical points and f(x) is increasing
throughout its domain. $f''(x) = -12(2 - 5x)^{-3}$
 $(-5) = 60/(2 - 5x)^3$, so there is no in-

flection point. f is concave upward on $(-\infty,2/5)$ and concave downward

29 (continued).

on $(2/5,\infty)$. There are no endpoints, local maxima, nor local minima.

Note that $x = 2/5$ is a vertical asymptote and $y = 3/5$ is a horizontal

asymptote.

33.   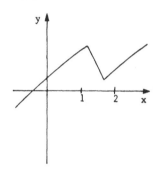   The statement is false. The figure shows

that the function may be decreasing within

the interval $(1,2)$.

37. This is true. Let $f(x) = ax^2 + bx + c$, $a \neq 0$. Then $f'(x) = 2ax +$

b and $f''(x) = 2a$, which never vanishes. Thus, parabolas don't have

inflection points.

41.   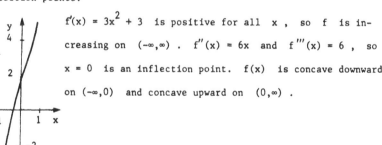 $f'(x) = 3x^2 + 3$ is positive for all x , so f is in-

creasing on $(-\infty,\infty)$. $f''(x) = 6x$ and $f'''(x) = 6$, so

$x = 0$ is an inflection point. $f(x)$ is concave downward

on $(-\infty,0)$ and concave upward on $(0,\infty)$.

45.   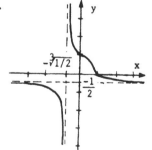   There is a vertical asymptote at $x = -\sqrt[3]{1/2}$.

$\lim\limits_{x\to\pm\infty} f(x) = -1/2$ is a horizontal asymptote.

$f'(x) = [(-3x^2)(1 + 2x^3) - (1 - x^3)(6x^2)] /$

$(1 + 2x^3)^2 = -9x^2/(1 + 2x^3)^2$, so $x = 0$ is

a critical point. $f''(x) = [-18x(1 + 2x^3)^2 -$

$(-9x^2)(2)(1 + 2x^3)(6x^2)]/(1 + 2x^3)^4 =$

45 (continued).

$18x(4x^3 - 1)/(1 + 2x^3)^3$, so possible inflection points are 0 and $\sqrt[3]{1/4}$. f'' changes sign at 0 and $\sqrt[3]{1/4}$, so these are inflection points. f is decreasing on its domain. It is concave upward on $(-\sqrt[3]{1/2},0)$ and $(\sqrt[3]{1/4},\infty)$; concave downward on $(-\infty,-\sqrt[3]{1/2})$ and $(0,\sqrt[3]{1/4})$.

49.

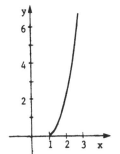

$y = x\sqrt{(x - 1)^3}$ implies $(x - 1)^3 \geqslant 0$, so the domain is $x \geqslant 1$. $f'(x) = (x - 1)^{3/2} + x(3/2) \times (x - 1)^{1/2} = (5x/2 - 1)(x - 1)^{1/2}$ implies the graph is strictly increasing. $f''(x) = 5(x - 1)^{1/2}/2 + (5x/2 - 1)(1/2)(x - 1)^{-1/2} = (15x - 2)/4\sqrt{x - 1}$ implies the graph is concave up.

53. Use the closed interval test of Section 3.5. $f'(x) = [(20x^3)/(x^2 + 1) - (5x^4 + 1)(2x)]/(x^2 + 1)^2 = 2x(5x^4 + 10x^2 - 1)/(x^2 + 1)^2$. Using the quadratic formula, we find the critical points to be $\pm[(-5 + \sqrt{30})/5]^{1/2}$ and 0 . The endpoints are -1 and 1 . $f(\pm1) = 3$; $f(0) = 1$; $f[\pm((-5 + \sqrt{30})/5)^{1/2}] = 2\sqrt{30} - 10 \approx 0.95$, so the maximum value is 3 and the minimum value is $2\sqrt{30} - 10$.

57.

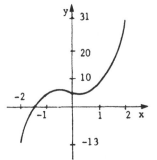

This function has no symmetry. $f'(x) = 9x^2 + 2x - 1$, which is zero at $x = (-2 \pm \sqrt{40})/18 = (-1 \pm \sqrt{10})/9$. $f''(x) = 18x + 2$, so the inflection point is at $x = -1/9$. At $x = (-1 - \sqrt{10})/9 \approx -0.46$, $f''(x) < 0$, so this is a local maximum. At $x = (-1 + \sqrt{10})/9 \approx 0.24$, $f''(x) > 0$,

so this is a local minimum.

61.

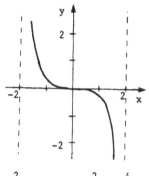

The graph is symmetric to the origin.

Vertical asymptotes exist at $x = \pm 2$.

$f'(x) = [(3x^2)(x^2 - 4) - x^3(2x)]/(x^2 - 4)^2 = x^2(x^2 - 12)/(x^2 - 4)^2$. The only critical point in $(-2,2)$ is $x = 0$, but $f'(x) \geq 0$ for all x in $(-2,2)$. $f''(x) = [(4x^3 - 24x)(x^2 - 4)^2 - (x^4 - 12x^2)(2) \times$

$(x^2 - 4)(2x)]/(x^2 - 4)^4 = 4x(2x^2 + 24)/(x^2 - 4)^3$, so the only possible inflection point is $x = 0$.

65.

Let r and h be the radius and height of the cone, and let R and H be the radius and height of the cylinder. We want to maximize $V = \pi R^2 H$. By similar triangles, we have

$R/(h - H) = r/h$, so $R = r(h - H)/h$, and $V = \pi r^2(h - H)^2 H/h^2$. Thus, $V'(H) = (\pi r^2/h^2)[2(h - H)(-1)H + (h - H)^2]$, and $V'(H) = 0$ implies $2H(h - H) = (h - H)^2$ or $2H = h - H$. Therefore, $H = h/3$ and $R = r(h - h/3)/h = 2r/3$.

69.

We are given that $V = \pi r^2 h$ and we want to minimize the cost, $C = (1/20)(2\pi rh + 2\pi r^2) + p(2)(2\pi r)$. $V = \pi r^2 h$ implies $h = V/\pi r^2$, so $C = (1/20)(2V/r + 2\pi r^2) + 4p\pi r = V/10r + \pi r^2/10 + 4p\pi r$. $C'(r) = -V/10r^2 + \pi r/5 + 4p\pi = (-V + 2\pi r^3 + 40p\pi r^2)/10r^2$. Therefore, the cheapest can has a radius given by $V = 2\pi r^3 + 40p\pi r^2$ and the height is given by $h = V/\pi r^2$, where the dimensions are given in centimeters.

73. (a) $T(x) = x^2(1 - x/4)$, $T'(x) = x(2 - 3x/4)$, $T'(2) = 1$.

(b)

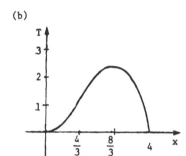

$T'(x) = 0$ at $x = 0, 8/3$. $T''(x) = 2 - 3x/2 > 0$ for $x < 4/3$ and $T''(x) < 0$ for $x > 4/3$, so T is concave upward on $[0,4/3)$ and concave downward on $(4/3,4]$. Also $T''(0) > 0$ and $T''(8/3) < 0$, so $x = 0$ is a local minimum and $x = 8/3$ is a local maximum. $T''(4/3) = 0$ and $T'''(4/3) \neq 0$, so 4/3 is an inflection point. $T(0) = T(4) = 0$; $T(8/3) = 64/27$; $T(4/3) = 32/27$.

77. (a) Revenue = price × quantity; total cost = fixed cost + cost of production where cost of production is cost per unit times number of units made. Profit = money taken in less the money used for production. Then substitute for R and C .

(b) $0 = (4.5)x - 9000$ has the solution $x = 2000$ units.

(c) $4500 = (4.5)x - 9000$ has the solution $x = 3000$ units.

(d) $P = R - C = [16x - (4/9)10^{-6}x^2] - [9000 + 11.5x] = -(4/9)10^{-6}x^2 + 4.5x - 9000$, so $P'(x) = -(8/9)10^{-6}x + 4.5$ and $P''(x) = -(8/9)10^{-6}$. $P'(x) = 0$ at $x = (81/16)10^6 = 5,625,000$. This maximizes profits since $P''(x) < 0$.

81. (a) $C'(x) = -2kx^{-3} + (-2)(3k)(40 - x)^{-3}(-1) = -2k(x^{-3} - 3(40 - x)^{-3}) = -2k[(40 - x)^3 - 3x^3]/x^3(40 - x)^3$. The critical points are found by solving $(40 - x)^3 = 3x^3$, which implies $(40 - x) = \sqrt[3]{3x}$ or $x = 40/(\sqrt[3]{3} + 1) \approx 16.4$. There are no other critical points.

(b) The critical point is the distance in miles from Smellter where particulate matter is minimized; therefore, the house should be built 16.4 miles from Smellter.

85. (a) 5 feet, 2 inches = 62 inches, so $W = (1/2)(62/10)^3 = 119.164$
pounds.

(b) $W'(h) = (3/2)(h/10)^2(1/10) = 3h^2/2000$, so $W'(48) = 432/125$ and
$W'(50) = 15/4$. By Consequence 1 of the mean value theorem, we
have $432/125 < [W(b) - W(a)]/(b - a) < 15/4$. Since $b - a = 2$,
$864/125 = 6.912 < W(50) - W(48) < 15/2 = 7.5$; therefore, the
child gains between 6.912 and 7.5 pounds. By direct calculation, $W(48) = 55.296$ and $W(50) = 62.5$, so $W(50) - W(48) =$
7.204 pounds.

89. (a)

(b)

$y = 10x(x - 1)^3(x - 3)^2$. For x
near 0 , $(x - 1)^3(x - 3)^2 \approx$
$(-1)^3(-3)^2 = -9$, so $y \approx -90x$.
For x near 1 , $10x(x - 3)^2 \approx$
$10(-2)^2 = 40$, so $y \approx 40(x - 1)^3$.
For x near 3 , $10x(x - 1)^3 \approx$
$30(2)^3 = 240$, so $y \approx 240(x - 3)^2$.

The graphs of these three functions give the shape of the graph of
the factored polynomial near its three roots (see the dotted curves).
Use the usual graphing procedure to complete the graph of y on
$[0,4]$.

93. As f is a polynomial of degree $n > 0$, f' is a polynomial of

    degree $n - 1 \geqslant 0$ with a finite number of roots in $(0,1)$ (at most

    $n - 1$), say $x_1 < x_2 < \ldots < x_{n-1}$. (Note: In the case of $n - 1 =$

    0, $f' = $ constant $\neq 0$ since f is nonconstant, which eliminated the

    only case where f' may have infinite number of roots.) As f' is

    continuous on $(0,1)$, f is strictly increasing or decreasing on each

    of the intervals between successive critical points. Suppose f has

    no local maxima or minima in $(0,1)$, then f' does not change sign

    in $(0,1)$. So f is increasing or decreasing on each of the intervals

    $(0,x_1)$, (x_1,x_2), $\ldots$, $(x_{n-1},1)$. By the continuity of f, f is

    increasing (decreasing) on each of the closed intervals $[0,x_1]$, $[x_1,x_2]$,

    $\ldots,[x_{n-1},1]$ and hence, on $[0,1]$. So $f(0) < f(1)$ or $f(0) > f(1)$,

    which contradicts the hypothesis $f(0) = f(1)$.

97. 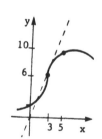 Let $g(x) = [f(x)]/x$, then the mean value

    theorem applied to $g(x)$ on $[3,5]$ gives us

    $(10/5 - 6/3)/(5 - 3) = 0$, so $g'(x_0) = 0$

    for some x_0 in $(3,5)$. We also know that

    $g'(x_0) = [x_0 f'(x_0) - f(x_0)]/x_0^2$, so $x_0 f'(x_0) =$

    $f(x_0) = y_0$ for some x_0 in $(3,5)$. This

    satisfies the general equation of a line passing through the origin,

    which is $y = xf'(x)$.

TEST FOR CHAPTER 3

1. True or false.

    (a) The intermediate value theorem tells you that $f(x) = 1/x$ has a
        root in $(-10,5)$ since $f(-10) < 0$ and $f(5) > 0$.

    (b) Any function that has a global minimum must have a local minimum.

    (c) A continuous function on a closed interval must have a local
        maximum.

    (d) If $f(x) = |3x|$, $f'(-1)$ does not exist.

    (e) If $f'(x_0) = 0$ and $f''(x_0) > 0$, then $f(x_0)$ is a local
        minimum.

2.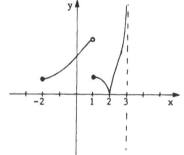

    (a) Is the figure shown at the left the
        graph of a function?

    (b) Where is the graph continuous?

    (c) Where is the function differentiable?

3. Give a graphical example to disprove this statement. "If $f(a) = \ell$,
    then $\lim_{x \to a} f(x)$ is always equal to ℓ ." Under what conditions is the
    statement true?

4. Show that the slope of the graph of $5x^3 + 9x^2 + 15x - 3$ must be 77
    somewhere in $(1,2)$ by two different methods.

    (a) Use the mean value theorem.

    (b) Use the intermediate value theorem.

5.   Use the six-step method to sketch the graph of $y = x^3/(x^2 + 1)$.

6.   Use the six-step method to sketch the graph of $y = (1 - x^2)/(x + 2)$.

7.   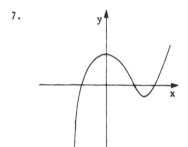 The sketch at the left depicts $f'(x)$. If $f(0) = 2$, make a sketch of the graph of $f(x)$.

8.   Find the maximum and minimum, if they exist, of $f(x) = 3x^3 - 3x^2 + 2$ on the following intervals:

(a) $[0,10]$

(b) $[-3,10)$

(c) $(-1,1)$

9.   Sketch the graph of $x^{2/3} + x^2$.

10. You're running out of money and you need to get back home. It's much cheaper to mail yourself than to fly home; however, you need to be shipped in a crate. The crate must be 6 feet high. A friend will give you 100 square feet of wood for the sides and bottom. Another material is used for the top. Maximize the volume of the crate.

(NOTE: People have tried this stunt in the past and suffocated.)

ANSWERS TO CHAPTER TEST

1. (a) False; there is a discontinuity in (-10,5) .

    (b) True

    (c) True

    (d) False; $f'(-1) = -3$

    (e) True

2. (a) Yes

    (b) (-2,1) and (1,3)

    (c) (-2,1) , (1,2) , and (2,3)

3. 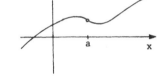 The statement is true for all continuous functions.

4. (a) $[f(2) - f(1)]/(2 - 1) = (103 - 26)/1 = 77.$
    (b) $f' = 15x^2 + 18x + 15$ is continuous; $f'(1) = 48 < 77 < 93 = f'(2)$.

5.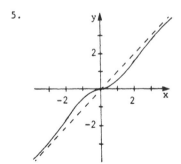

6.

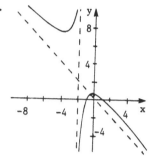

7.

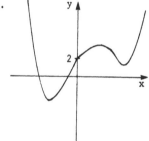

8.   (a)   Maximum = 2702 ;   minimum = 14/9 .

   (b)   No maximum; minimum = -106 .

   (c)   Maximum = 2 ;   no minimum.

9.

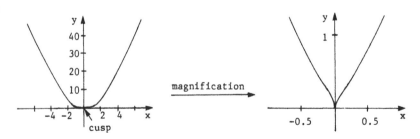

10.   Dimensions = $(-12 + 2\sqrt{61}) \times (-12 + 2\sqrt{61}) \times 6$;

   volume = $(2328 - 288\sqrt{61})$ cubic feet.

COMPREHENSIVE TEST FOR CHAPTERS 1-3 (Time limit: 3 hours)

1.   True or false.   If false, explain why.

(a) The slope of f(x) at (1,f(1)) is f'(1) .

(b) (d/dx)[(3x^2 + x)(2x^4 + x)] = (6x + 1)(8x^3 + 1) .

(c) Suppose that f(x) = π^3 + x^4 - 2x^2 + 1 , then f'(x) = 3π^2 + 4x^3 - 4x .

(d) If (a,f(a)) is the maximum of a differentiable function on (-∞,∞) , then f'(a) can only be zero.

(e) Suppose that. u = h(z) = z^3 - 4z , then the tangent line at z = 2 is u = (3z^2 - 4)(z - 2) .

(f) The most general antiderivative of (5x^2 + 3)(25x^2 + 3) is x(5x^2 + 3)^2 + C .

(g) A local minimum of f(x) may have a value larger than a local maximum of f(x) .

(h) All continuous functions are differentiable.

(i) The product of two increasing functions is also an increasing function.

2.   This question refers to f(x) = (x^2 - x + 3)/(x + 2).

(a) Is f(x) symmetric to the x-axis, y-axis, origin, or neither?

(b) Where are the asymptotes of f(x) ?

(c) Where does f(x) = 0 ?

(d) On what intervals is f(x) increasing? decreasing?

(e) On what intervals is f(x) concave upward? concave downward?

(f) Where are the inflection points?

(g) Sketch the graph of f(x) .

3. Short answer.

   (a) Express $f'(x)$ as a limit.

   (b) If $g(x) = \sqrt{x^5 - 1}$, what is d^2g/dx^2 ?

   (c) Sketch the parametric curve $x = t$, $y = \sqrt{t}$.

   (d) Define an inflection point.

   (e) Approximate $(1.9)^2$.

4. Multiple choice.

   (a) The derivative of $1/(-x^2 + 2)^3$ is:

   (i) $1/(-6x)(2 - x)^2$ (ii) $6x/(-x^2 + 2)^4$

   (iii) $6x/(2 - x^2)^2$ (iv) $-2x/(-x^2 + 2)^4$

   (b) If $x^2 + y^2 = 1$, then dy/dx is:

   (i) $-y/x$ (ii) $-x/y$

   (iii) $2x$ (iv) $2y$

   (c) If $h'(v) > 0$, then $h(v)$ is

   (i) increasing   (ii) decreasing

   (iii) positive   (iv) negative

   (d) $f(x) = x^3 + 2x + 1$ crosses the x-axis once in the interval:

   (i) $(-2,-1)$ (ii) $(-1,0)$

   (iii) $(0,1)$ (iv) $(1,2)$

   (e) The mean value theorem states that:

   (i) the average value of $f(a)$ and $f(b)$ is attained in (a,b)
   if f is continuous.

   (ii) all continuous functions have a mean.

   (iii) all differentiable functions have a mean.

   (iv) the average slope of f in $[a,b]$ is in (a,b) if f is
   differentiable.

5. Differentiation problems.

   (a) Find dy/dx if $y = (x^2 - 3)\sqrt{1/2 - x}$.

   (b) Find dy/dx if $y = (x^{2/3} + 2/3)/(x + \sqrt{x})$.

   (c) Find dy/dx if $y = 3t$ and $x = 4t^2 - t$.

   (d) Find dy/dx if $x^2 y^{1/3} = xy$.

   (e) Find dy/dx if $y = f(g(x))$ where $f(x) = \sqrt{x^2 + x}$ and $g(x) = x^3 + x^{3/2}$.

6. Find the equation of the cubic polynomial $f(x)$ whose graph has an inflection point at $x = 2$ and a local maximum at $(0,5)$.

7. Antidifferentiation problems.

   (a) Evaluate $\int (\sqrt{t} + t^{-2} + 1)dt$.

   (b) Evaluate $\int \sqrt{5y/3 + 3/5}\ dy$.

   (c) Show that an antiderivative of $f(x) = (4x + 6)/(x^2 + 3x + 1)^2$ is $-2(x^2 + 3x + 1)^{-1} + 6$.

8. Water is being pumped from a square swimming pool with a length of 10 m. into a circular cylinder with radius 3 m. How fast is the water level rising in the cylinder if the water in the pool is dropping 10 cm./min. ?

9. (a) The difference between x and y is 10 . Find the minimum of xy .

   (b) Suppose x and y are related by $5x^2 + y = 5$. Find the maximum of $x^2 y$.

10. At time t , a particle's position is given by $y = x^4/4 - 3x^2/2 + 4x$. Sketch the graph of the particle's velocity function.

ANSWERS TO COMPREHENSIVE TEST

1. (a) True.

   (b) False; the derivative is $(6x + 1)(2x^4 + x) + (3x^2 + x)(8x^3 + 1)$.

   (c) False; π^3 is a constant whose derivative is zero.

   (d) True.

   (e) False; the tangent line is $u = (3z^2 - 4|_2(z - 2) = 8(z - 2)$.

   (f) True.

   (g) True.

   (h) False; consider $f(x) = |x|$.

   (i) False; let $f(x) = g(x) = x$ on $[-1,0]$.

2. (a) Neither.

   (b) Vertical asymptote: $x = -2$; no horizontal asymptote.

   (c) Nowhere.

   (d) Increasing: $(-\infty,-5)$ and $(1,\infty)$; decreasing: $(-5,-2)$ and $(-2,1)$.

   (e) Concave upward: $(-2,\infty)$; concave downward: $(-\infty,-2)$.

   (f) None.

   (g)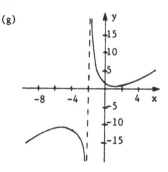

3. (a) $\lim\limits_{\Delta x \to 0} \{[f(x + \Delta x) - f(x)]/\Delta x\}$
   (b) $5x^3(3x^5 - 8)/4(x^5 - 1)^{3/2}$

3 (continued)

(c)

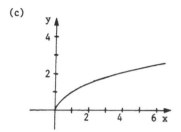

(d) An inflection point is the point where concavity changes direction.

(e) 3.6

4. (a) ii

(b) ii

(c) i

(d) ii

(e) iv

5. (a) $(2x)\sqrt{1/2 - x} - (x^2 - 3)/2\sqrt{1/2 - x}$

(b) $[(2/3x^{1/3})(x + \sqrt{x}) - (x^{2/3} + 2/3)(1 + 1/2\sqrt{x})]/(x + \sqrt{x})^2$

(c) $3/(8t - 1)$

(d) $(2xy^{1/3} - y)/(x - x^2/3y^{2/3})$

(e) $[2(x^3 + x^{3/2}) + 1](3x^2 + 3\sqrt{x}/2)/2[(x^3 + x^{3/2})^2 + (x^3 + x^{3/2})]^{1/2}$

6. $f(x) = x^3/8 - 3x^2/4 + 5$

7. (a) $2t^{3/2}/3 - 1/t + t + C$

(b) $(2/5)(5y/3 + 3/5)^{3/2} + C$

(c) Differentiate $2(x^2 + 3x + 1)^{-1} + 6$ to get $f(x)$.

8. $(1000/9\pi)$ cm./min.

9. (a) -25

(b) $5/4$

10.

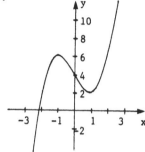

CHAPTER 4

THE INTEGRAL

4.1 Summation

PREREQUISITES

1. There are no prerequisites other than simple addition and algebra for this section.

GOALS

1. Be able to manipulate expressions involving summation notation.

STUDY HINTS

1. Dummy index. Changing the index letter does not change the value of the sum. Understand this concept.

2. Properties of summation. With use, you will soon have these properties memorized, so there is no need to actively memorize these properties. However, you should understand the meaning of each statement. Note how

$$\sum_{i=1}^{n} (a_i + b_i) = \sum_{i=1}^{n} a_i + \sum_{i=1}^{n} b_i \quad \text{is analogous to} \quad \int (f + g)dx = \int f \, dx + \int g \, dx ,$$

which was introduced in Section 2.5 (See p. 130).

3. Sums of products. Be cautious that the following statement does not hold in general: $\sum_{i=m}^{n} a_i b_i \overset{?}{=} \left(\sum_{i=m}^{n} a_i \right) \left(\sum_{i=m}^{n} b_i \right) .$

4.   <u>Sum of the first n numbers</u>. Memorize or learn how to derive this formula.

5.   <u>Substitution of index</u>.   Formula (6) should be understood and not memorized.   Note that the left-hand side of the formula in the text begins at a_{m+q} and ends at a_{n+q} .   Be sure the sums begin and end with the same terms.   For example, if $\sum_{j=1}^{5} 2^{j+2} = \sum_{i=x}^{y} 2^{i-3}$, then x = 6 and y = 10 because the exponents of 2 on both sides should begin at 3 and end at 7 .   (Note that the indices do not always start at 1 or 0 .)

6.   <u>Telescoping sums</u>.   There are several other ways for writing the telescoping sum, such as $\sum_{i=m}^{n} [a_i - a_{i-1}] = a_n - a_{m-1}$.   These formulas can easily be recognized by their minus signs and in many cases, a shift of the index by one .   Again, don't memorize these formulas.   To compute a telescoping sum, write down a few terms at the beginning and at the end of the sum, and then cancel out the terms in the middle.

Try Example 7 with $(i - 1)^3$ replaced by $(i + 3)^3$.   This new example demonstrates that your answer may have more than two terms.

SOLUTIONS TO EVERY OTHER ODD EXERCISE

1.   Use the formula $\Delta d = \sum_{i=1}^{n} v_i \Delta t_i$ to get $\Delta d = 2(3) + 1.8(2) + 2.1(3) + 3(1.5) = 20.4$ meters.

5. $\sum_{i=1}^{4} (i^2 + 1) = (1^2 + 1) + (2^2 + 1) + (3^2 + 1) + (4^2 + 1) = 2 + 5 + 10 + 17 = 34$.

9. Use the formula for summing the first n integers to get $1 + 2 + \ldots +$

$$25 = \sum_{i=1}^{25} i = (1/2)(25)(26) = 325 .$$

13. Use the method of Example 6 and then use the formula for the sum of the

first n integers to get $\sum_{j=4}^{80} (j - 3) = \sum_{i=1}^{77} i = (1/2)(77)(78) = 3003 .$

17. Note that the nth term cancels the $-(n^{th})$ term, so $\sum_{j=-2}^{2} j^3 = 0 .$

21. Using property 3 of summation, we have $\sum_{i=1}^{n} i = \sum_{i=1}^{m-1} i + \sum_{i=m}^{n} i .$ Rearrange

the equation and use the formula for the sum of the first n integers

to get $\sum_{i=m}^{n} i = \sum_{i=1}^{n} i - \sum_{i=1}^{m} i = (1/2)(n)(n + 1) - (1/2)(m - 1)(m) .$

25. Use the method of Example 7 to compute the telescoping sum. $\sum_{i=1}^{100} [i^4 -$

$(i - 1)^4] = [(1)^4 - (0)^4] + [(2)^4 - (1)^4] + \ldots + [(99)^4 - (98)^4] +$

$[(100)^4 - (99)^4] = (100)^4 - (0)^4 = 100,000,000 .$

29.

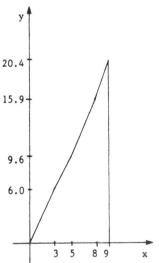

Graphically, the velocity is the slope.

33. This telescoping sum is $\sum\limits_{k=1}^{n} [(k+1)^4 - k^4] = [(2)^4 - (1)^4] +$

$[(3)^4 - (2)^4] + \ldots + [(n-1+1)^4 - (n-1)^4] + [(n+1)^4 - n^4] =$

$(n+1)^4 - 1$.

37. $\sum\limits_{i=-30}^{30} (i^5 + i + 2) = \sum\limits_{i=-30}^{30} (i^5 + i) + \sum\limits_{i=-30}^{30} 2$. Notice that the nth

term cancels the $-(n^{th})$ term in $\sum\limits_{i=-30}^{30} (i^5 + i)$. Using Property 4 of

summation, the sum is $0 + 2(61) = 122$.

41. (a) Rearrange the given equation and sum both sides. $3 \sum\limits_{i=1}^{n} i^2 =$

$\sum\limits_{i=1}^{n} [(i+1)^3 - i^3] - 3 \sum\limits_{i=1}^{n} i - \sum\limits_{i=1}^{n} 1 = [(n+1)^3 - 1] -$

$3[n(n+1)/2] - n = [n^3 + 3n^2 + 3n] - [(3n^2 + 3n)/2] - n = n^3 +$

$(3/2)n^2 + (1/2)n$. Dividing both sides by 3 , we get

$\sum\limits_{i=1}^{n} i^2 = n^3/3 + n^2/2 + n/6 = [n(n+1)(2n+1)]/6$.

(b) If m is positive, then $\sum\limits_{i=m}^{n} i^2 = \sum\limits_{i=0}^{n} i^2 - \sum\limits_{i=0}^{m-1} i^2 =$

$[n(n+1)(2n+1)]/6 - [(m-1)m(2m-1)]/6$.

If m is 0 , then $\sum\limits_{i=0}^{n} i^2 = [n(n+1)(2n+1)]/6$ and the

previous result still holds.

If m is negative, the $\sum\limits_{i=m}^{n} i^2 = \sum\limits_{i=0}^{n} i^2 + \sum\limits_{i=m}^{0} i^2 =$

$[n(n+1)(2n+1)]/6 + [m(m+1)(2m+1)]/6$. However, we get the

same result by substituting $-m$ for m in $[n(n+1)(2n+1)]/6 -$

$[(m-1)m(2m-1)]/6$; therefore, $\sum\limits_{i=m}^{n} i^2 = [n(n+1)(2n+1)]/6 -$

41. (b) continued.

$$[(m - 1)m(2m - 1)]/6 = (1/6)[2(n^3 - m^3) + 3(n^2 + m^2) + (n - m)],$$

regardless of the sign of m .

(c) $\sum_{i=1}^{n} [(i + 1)^4 - i^4] = (n + 1)^4 - 1$ because this is a telescoping

sum. $(i + 1)^4 - i^4 = 4i^3 + 6i^2 + 4i + 1$. Rearrangement and

summing yields $4 \sum_{i=1}^{n} i^3 = \sum_{i=1}^{n} [(i + 1)^4 - i^4] - 6 \sum_{i=1}^{n} i^2 - 4 \sum_{i=1}^{n} i -$

$\sum_{i=1}^{n} 1 = [(n + 1)^4 - 1] - [n(n + 1)(2n + 1)] - 2n(n + 1) - n =$

$n^2(n^2 + 2n + 1)$. Thus, $\sum_{i=1}^{n} i^3 = [n(n + 1)/2]^2$.

SECTION QUIZ

1. Compute $\sum_{i=1}^{n} 2 + \sum_{i=-5}^{m} i + \sum_{j=0}^{n} 3j$. Assume that m and n are positive

integers.

2. Find x and y , and then compute the resulting sum: $\sum_{j=-4}^{45} [(j + 6)^2 -$

$(j + 4)^2] = \sum_{k=x}^{y} [(k + 2)^2 - k^2]$.

3. Find $\sum_{i=0}^{3} 3i$ and $\left(\sum_{i=0}^{3} 3 \right) \left(\sum_{i=0}^{3} i \right)$. What fact does your answer demonstrate?

4. (a) Consider $\sum_{i=2}^{2} i$. Is this expression defined? If so, what is its

value?

(b) Consider $\sum_{i=2}^{1} i$. Is this expression defined? If so, what is its

value?

5. Your rich, eccentric uncle's will requests that you drive his ashes
    around town in his Rolls Royce. He offers to pay you 7000j dollars
    on the jth day of the month. On the other hand, he will pay you
    2^{j+1} cents for this service on the jth day provided you return
    2^{j-1} cents. If you work during April, how much could you earn by
    each form of payment? (Note: 2^{30} = 1,073,741,824.)

ANSWERS TO SECTION QUIZ

1. $2n + m(m + 1)/2 - 15 + 3n(n + 1)/2$.

2. $x = 0$, $y = 49$;   sum $= (51)^2 + (50)^2 - (1)^2 - (0)^2 = 5100$.

3. $\sum_{i=0}^{3} 3i = 18$; $\left(\sum_{i=0}^{3} 3 \right)\left(\sum_{i=0}^{3} i \right) = 12(6) = 72$.   This shows that the sum of

     products does not equal the product of the sums of the multiplicands.

4.   (a) Yes. $\sum_{i=2}^{2} i = 2$.

     (b) No.

5. $3,255,000 by the first method; $32,212,254.69 by the second method.

4.2 Sums and Areas

PREREQUISITES

1. Recall how to manipulate expressions involving the summation notation
    (Section 4.1).

PREREQUISITE QUIZ

1. Fill in the blank: $\displaystyle\sum_{i=1}^{n} (i + 3) = $ _____ $+ \displaystyle\sum_{i=1}^{n} i$.

2. Compute the sum in Question 1 .

3. Consider $\displaystyle\sum_{j=1}^{4} x_j y_j$. Let $x_j = 2^j$ and $y_j = j - 2$. Compute the sum.

GOALS

1. Be able to find x_i , Δx_i , and k_i for a given step function, and be
    able to find the area of the region under its graph.

2. Be able to state the relationship between upper sums, lower sums, and
    the area under a positive function.

STUDY HINTS

1. Area under graphs. The area under a graph is closely related to the
    important concept of the integral, so be sure you know exactly what the
    boundaries of the area are. Note that one or both of a and b in
    Fig. 4.2.2 could be negative.

2. Step functions. Be able to define a step function. In addition, be
    familiar with the notations x_i , Δx_i , and k_i . Note that the first
    x is x_0 , not x_1 . This is because there are $n + 1$ partition
    points and only n intervals.

3.   <u>Upper and lower sums</u>. Note that a lower sum is the area of <u>any</u> step

function which lies entirely within the region under a graph. Thus,

there are infinitely many lower sums. Also, note that the subintervals

may have different widths. Similar statements may be made for upper sums.

4.   <u>Relationship between upper sums, lower sums, and area</u>. Know the following

inequality: lower sums $\leqslant$ area under a graph $\leqslant$ upper sums.

5.   <u>Direct calculation of areas</u>. Example 6 shows how areas were computed

prior to the invention of calculus. Do not be overly concerned with

understanding this example. You will be given a simple formula for

computing areas in Section 4.4 .

SOLUTIONS TO EVERY OTHER ODD EXERCISE

1.   As in Example 1, the graph consists of horizontal

lines with heights 0 , 2 and 1 on their respec-

tive intervals. Solid dots are used to indicate

that the function includes the endpoint. Open

dots indicate that the endpoint is not to be included.

5. x_0 is the first endpoint. The other x_i's occur where the function

changes value or at the last endpoint. Thus, $x_0 = 0$, $x_1 = 1$, $x_2 = 2$, and $x_3 = 3$. By definition, $\Delta x_i = x_i - x_{i-1}$; so $\Delta x_1 = x_1 - x_0 = 1 - 0 = 1$. Similarly, $\Delta x_2 = \Delta x_3 = 1$. The k_i's are the func-

tion values on the ith interval, so $k_1 = 0$, $k_2 = 2$, and $k_3 = 1$.

The area under the graph is $\sum_{i=1}^{3} k_i \Delta x_i = 0(1) + 2(1) + 1(1) = 3$.

9.

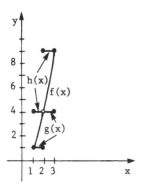

The upper sum is the area under h(x) ,

shown as a dotted line. Its area is

$$\sum_{j=1}^{2} \ell_j \Delta x_j = 4(1) + 9(1) = 13 .$$ The

lower sum is the area under g(x) ,

shown as horizontal solid lines. The

area is $\sum_{i=1}^{2} k_i \Delta x_i = 1(1) + 4(1) = 5$.

13. The problem is analogous to Example 5. According to Exercise 9, the

lower and upper sums are 5 and 13 , so the distance crawled is be-

tween 0.010 and 0.026 meters, i.e., between 10 and 26 millimeters.

17. Using the method of Example 6, partition [a,b] into n equal subinter-

vals. Thus, the partition is (a, a + (b - a)/n , a + 2(b - a)/n , ... ,

a + (n - 1)(b - a)/n, b) . We will find upper and lower sums and use the

fact that $L \leqslant A \leqslant U$. On the interval (a + (i - 1)(b - a)/n , a +

i(b - a)/n) , let g(x) = 5[a + (i - 1)(b - a)/n] and let h(x) =

5[a + i(b - a)/n] . Therefore, a lower sum is $\sum_{i=1}^{n} 5[a + (i - 1)(b - a)/$

n][(b - a)/n] because $\Delta x_i = (b - a)/n$. This is $[5a(b - a)/n] \sum_{i=1}^{n} 1 +$

$[5(b - a)^2/n^2] \sum_{i=0}^{n-1} i = 5a(b - a) + 5(b - a)^2(n - 1)/2n = 5(ab - a^2) +$

$5(b^2 - 2ab + a^2)/2 - 5(b - a)^2/2n = 5(b^2 - a^2)/2 - 5(b - a)^2/2n$.

Similarly, an upper sum is $\sum_{i=1}^{n} 5[a + i(b - a)/n][(b - a)/n] =$

$[5a(b - a)/n] \sum_{i=1}^{n} 1 + [5(b - a)^2/n^2] \sum_{i=1}^{n} i = 5a(b - a) + 5(b - a)^2(n + 1)/2n =$

$5(b^2 - a^2)/2 + 5(b - a)^2/2n$. Since $L \leqslant A \leqslant U$ for all n , we must

have $A = 5(b^2 - a^2)/2$.

21.   According to the result of Exercise 15, the area   x   on $[1,2]$ is   3/2 .
By the result of Exercise 20, the area under x^2 on $[0,1)$ is
$(1/3)(1^3 - 0^3) = 1/3$.   Finally, by the additive property for areas, the
area under $f(x)$ on $[0,2]$ is   3/2 + 1/3 = 11/6 .

SECTION QUIZ

1.   True or false: For a given partition, x_0 may be negative.

2.   True or false: If you know Δx_1 ,   then all of the other Δx_i must
equal Δx_1 .

3.   For a given non-negative function, can a lower sum ever equal an upper
sum?   Explain your answer.

4.   (a) y   (b) y   (c) y

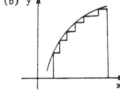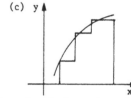

In each case sketched, the area under the step function is less than the
area under $f(x)$.   Which, if any, of the step functions is a lower sum?

5.   Your new office building was constructed by a carpenter who enjoyed
cocktail lunches.   Upon completion, you discover that the drunk has
made one of the walls curved.   Measured in feet, the wall follows the
curve $y = x^3 + 6x^2 + 3$ on $[-5,3]$.

   (a)   The floor is to be lined with one-foot square tiles which cannot be
         cut.   The sides of the tiles are placed parallel to the coordinate
         axes.   How much floor area may be tiled?

   (b)   Upon closer inspection, the curved wall has a crack on the bottom
         which permits tile to be slipped under the wall.   How many tiles
         are needed to completely tile the floor?

5.   (c)   From your answers in parts (a) and (b), what do you know about the exact area of the floor?

ANSWERS TO PREREQUISITE QUIZ

1. $\sum\limits_{i=1}^{n} 3 = 3n$

2. $n^2/2 + 7n/2$

3.   38

ANSWERS TO SECTION QUIZ

1.   True.

2.   False. Δx_i may have any positive length.

3.   They are equal only when the number is precisely the area under the curve.

4.   (a) and (b).   (c) does not satisfy $g(x) \leqslant f(x)$ on $[a,b]$.

5.   (a)   136 square feet.

     (b)   256 tiles.

     (c)   136 sq. ft. $\leqslant$ area $\leqslant$ 256 sq. ft.

4.3 Definition of the Integral

PREREQUISITES

1.   Recall how step functions are used to determine the area under a positive
     function (Section 4.2).

2.   Recall the definitions for upper and lower sums (Section 4.2).

PREREQUISITE QUIZ

1.   Define upper and lower sums. Give summation formulas for computing them.

2.   How are upper and lower sums related to the area under the graph of a
     positive function?

GOALS

1.   Be able to use the concepts of step functions and upper and lower sums
     from Section 4.2 to estimate the signed area under a general function.

2.   Be able to relate signed area to the integral.

3.   Be able to write integrals as a Riemann sum.

STUDY HINTS

1.   Signed area. Know that signed area means that the area below the x-axis
     is subtracted from the area above the x-axis. The computational formula
     is the same as the one used for area in the last section.

2.   Upper sum, lower sum, and area. The definitions and the formulas for
     upper and lower sums are the same as those in Section 4.2 . Also, upper
     and lower sums are related to signed areas in exactly the same way they
     were related to areas in the previous section.

3.   **Integrals and signed areas.** It is important to know that the integral
     and the signed area under a curve are equal.

4.   **Definition of the integral.** We will define the integral to be the unique
     number which lies between all lower sums and all upper sums.

5.   **Estimating integrals.** Referring to Example 4, note that estimates
     usually use equal subdivisions. In general, more subdivisions increase
     the accuracy of the estimate. Also note that for functions which are
     strictly increasing or decreasing, your upper and lower sums will differ
     only in the first and last terms. This is because the estimate uses a
     telescoping sum. Thus, the difference is $|f(x_0) - f(x_n)|\Delta x$. When
     estimating integrals of general functions, be sure to consider the
     critical points when you compute upper and lower sums. Why consider
     the critical points? Try to estimate a lower sum for $f(x) = x^2 - x$
     on $[0,1]$ with $n = 1$.

6.   **Integrability versus differentiability.** Non-continuous functions may
     be integrated whereas they cannot be differentiated. For example, a
     step function may be integrated over an interval, but it is not
     differentiable.

7.   **Riemann sums.** Note that c_i may be chosen as _any_ point in $[x_{i-1}, x_i]$.
     Note also that the definition requires the number of subdivisions to go
     to ∞ and that the largest subdivision approach zero.

8.   **Physical motivation.** Although the supplement may be skipped, it will
     give you an appreciation of the usefulness and the practicality of the
     integral.

SOLUTIONS TO EVERY OTHER ODD EXERCISE

1.   The signed area of the step function depicted at

the left is $\sum\limits_{i=1}^{2} k_i \Delta x_i = (1)(1 - 0) + (-3)(2 - 1) = -2$.

5. $\int_a^b g(x)\,dx$ is the signed area of the region between $g(x)$ and the x-axis from $x = a$ to $x = b$. From the solution to Exercise 1, we know that $\int_0^2 g(x)\,dx = -2$.

9.   Since the signed area equals the integral, the answer is $\int_{-1}^{2} x^2\,dx$.

13. Applying the method of Example 4, we use the partition $(2, 8/3, 10/3, 4)$.

Thus, an upper sum is $[1/2 + 1/(8/3) + 1/(10/3)](2/3) = 37/60$.

Similarly, a lower sum is $[1/(8/3) + 1/(10/3) + 1/4](2/3) = 47/60$.

The integral lies between $37/60$ and $47/60$, a difference of

$1/6$. Therefore, $42/60 = 7/10$ must be within $1/12$ of $\int_2^4 (dx/x)$.

A better estimate may be obtained by using smaller subdivisions.

17. This problem is analogous to Example 5.

(a) The displacement of the bus is $\int_0^3 5(t^2 - 5t + 6)\,dt$.

(b) Note that $v = 5(t - 3)(t - 2)$, so it is positive on $(0,2)$ and negative on $(2,3)$. Thus, the total distance travelled is $\int_0^2 5(t^2 - 5t + 6)\,dt - \int_2^3 5(t^2 - 5t + 6)\,dt$.

21. Divide the interval into n equal parts to get the partition

$(2, 2 + 2/n, 2 + 4/n, \ldots, 2 + 2(n - 1)/n, 4)$. Choose $c_i = 2 + 2i/n$, so $S_n = \sum\limits_{i=1}^{n} \{1/[1 + (2 + 2i/n)^2]\}(2/n) = \sum\limits_{i=1}^{n} [n^2/(n^2 + 4n^2 + 8in + 4i^2)](2/n) = \sum\limits_{i=1}^{n} [2n/(n^2 + 4n^2 + 8in + 4i^2)]$. Therefore,

21. (continued)

$$\lim_{n \to \infty} \sum_{i=1}^{n} [2n/(5n^2 + 8in + 4i^2)] = \int_2^4 [dx/(1 + x^2)] .$$ (Answers may vary

depending on the choice of c_i.)

25. (a) If $0 \leqslant x < 1$, we can take $(0,x)$ as our partition and

$\int_0^x f(t)dt = 2(x - 0) = 2x$. If $1 \leqslant x < 3$, we take $(0,1,x)$ as

our partition and $\int_0^x f(t)dt = 2 \cdot (1 - 0) + 0 \cdot (x - 1) = 2$. If

$3 \leqslant x \leqslant 4$, we take $(0,1,3,x)$ as our partition and $\int_0^x f(t)dt =$

$2(1 - 0) + 0(3 - 1) + (-1)(x - 3) = 2 - x + 3 = -x + 5$.

Summarizing, we have
$$\int_0^x f(t)dt = \begin{cases} 2x & \text{if } 0 \leqslant x < 1 \\ 2 & \text{if } 1 \leqslant x < 3 \\ -x + 5 & \text{if } 3 \leqslant x \leqslant 4 \end{cases}.$$

(b)

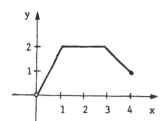

(c) We can see from the graph that F is differentiable everywhere on

(0,4] except at the points 1 , 3 , and 4 . We have:

$$F'(x) = \begin{cases} 2 & \text{if } 0 < x < 1 \\ 0 & \text{if } 1 < x < 3 \\ -1 & \text{if } 3 < x < 4 \end{cases}.$$

We see that F' is the same as f , except at the points where the

piecewise constant function f has a jump.

29. (a) The total volume is the sum of the volumes of the individual pieces,

$$\text{i.e.,}\quad \sum_{i=1}^{n} A_i \Delta x_i .$$

(b) Consider a graph of cross-sectional area, $f(x)$, versus the distance, x , from the end of the rod. The volume is the area under the curve or $\int_0^L f(x)\,dx$ (assuming f to be integrable).

33. (a)

$$f(x) + g(x) = \begin{cases} 6 & 0 \leqslant x < 1 \\ 1 & 1 \leqslant x < 3/2 \\ 0 & 3/2 \leqslant x < 2 \\ 3 & 2 \leqslant x \leqslant 3 \end{cases}.$$

This is a step function. Use the partition $(0 , 1 , 3/2 , 2 , 3)$.

Therefore, $\int_0^3 [f(x) + g(x)]\,dx = 6(1 - 0) + 1(3/2 - 1) + 0(2 - 3/2) + 3(3 - 2) = 19/2$.

(b) Use the partition, $(1 , 3/2 , 2)$, so $\int_1^2 [f(x) + g(x)]\,dx = 1(3/2 - 1) + 0(2 - 3/2) = 1/2$.

(c) $\int_0^3 2f(x)\,dx = 12(1 - 0) + 2(3/2 - 1) + 0(2 - 3/2) + 6(3 - 2) = 19$; $2\int_0^3 f(x)\,dx = 2(19/2) = 19 = \int_0^3 2f(x)\,dx$.

(d)

$$f(x) - g(x) = \begin{cases} 2 & 0 \leqslant x < 1 \\ -3 & 1 \leqslant x < 3/2 \\ -2 & 3/2 \leqslant x < 2 \\ 1 & 2 \leqslant x \leqslant 3 \end{cases}.$$

Using the same partition as in part (a), $\int_0^3 [f(x) - g(x)]\,dx = 2(1 - 0) + (-3)(3/2 - 1) + (-2)(2 - 3/2) + 1(3 - 2) = 1/2$.

(Continued on next page.)

33.   (d) (continued)

$\int_0^3 f(x)dx - \int_0^3 g(x)dx = [(4)(1 - 0) + (-1)(3/2 - 1) + (-1)(2 - 3/2) +$

$(2)(3 - 2)] - [(2)(1 - 0) + (2)(3/2 - 1) + (1)(2 - 3/2) +$

$(1)(3 - 2)] = 5 - 9/2 = 1/2$.   Therefore, $\int_0^3 [f(x) - g(x)]dx =$

$\int_0^3 f(x)dx - \int_0^3 g(x)dx$.

(e)

$$f(x) \cdot g(x) = \begin{cases} 8 & 0 \leqslant x < 1 \\ -2 & 1 \leqslant x < 3/2 \\ -1 & 3/2 \leqslant x < 2 \\ 2 & 2 \leqslant x \leqslant 3 \end{cases} .$$

$\int_0^3 f(x) \cdot g(x)dx = 8(1 - 0) + (-2)(3/2 - 1) + (-1)(2 - 3/2) + 2(3 - 2) =$

15/2 .   From part (d), we have $\int_0^3 f(x)dx \cdot \int_0^3 g(x)dx = 5(9/2) = 45/2$;

therefore, the statement is false.

SECTION QUIZ

1.   State whether the following statements are true or false.

(a)   There is no such thing as a negative area.

(b)   Any integrable function is differentiable.

(c)   Any differentiable function is integrable.

2.   Circle the appropriate word: The integral is (exactly, approximately,
     not) equal to the signed area under the curve.

3.   You have purchased a strip of land from the Rolling Hills Real Estate
     Company.   Along a straight line, the land has altitude $x^3 + 2x^2 - 4x -$
     3 , where the x-axis represents sea level.   Measurements were made in
     kilometers.   Your pet buffaloes like to roam over level ground, so soil
     from the hills is used to fill in the valleys.

3.   (a) Using the partition $(-3 , -2 , -1 , 0 , 1 , 2 , 3)$,   find an
          upper sum and a lower sum to estimate the height of the ground on
          the interval $[-3,3]$.

     (b) Write the exact height as a limit.

ANSWERS TO PREREQUISITE QUIZ

1.   An upper sum for a function $f(x)$ is the area under a step function
     $h(x)$ for which $h(x) \geqslant f(x)$ on the interval of definition. A lower
     sum is the area under a step function $g(x)$ for which $g(x) \leqslant f(x)$.

     The formulas are $U = \sum_{j=1}^{m} \ell_j \Delta x_j$ and $L = \sum_{i=1}^{n} k_i \Delta x_i$.

2.   Lower sums $\leqslant$ Area $\leqslant$ Upper sums.

ANSWERS TO SECTION QUIZ

1.   (a) False. This involves the concept of signed areas.

     (b) False. Consider the partition points of a step function.

     (c) True. All differentiable functions are continuous.

2.   Exactly.

3.   (a) Since the strip of land is 6 kilometers long, the upper sum is
          $44/6 = 22/3$ kilometers; Lower sum $= (-256/27)/6 = -128/51$ kilometers.
          (Did you consider the critical point at $x = 2/3$?)

     (b) $\displaystyle \lim_{n \to \infty} [(396/n^2) \sum_{i=1}^{n} i - (1452/n^3) \sum_{i=1}^{n} i^2 + (1296/n^4) \sum_{i=1}^{n} i^3]$.

4.4   The Fundamental Theorem of Calculus

PREREQUISITES

1.   Recall how to compute antiderivatives using the power rule, the sum
     rule, and the constant multiple rule (Section 1.4).

PREREQUISITE QUIZ

1.   Find an antiderivative for the following functions:

     (a) x

     (b) x^4

     (c) x^n , n = integer, $\neq -1$.

2.   Use the constant multiple rule to find an antiderivative for:

     (a) $3x$

     (b) $49x^6$

     (c) $(n + 1)x^n/2$,   where   n = integer, $\neq -1$

     (d)   5

3.   Find an antiderivative for:

     (a) $x + 3$

     (b) $x^4 + 3x + 5$

GOALS

1.   Be able to state and apply the fundamental theorem of calculus.

2.   Be able to evaluate integrals by using antiderivatives.

STUDY HINTS

1.   <u>Fundamental theorem of calculus</u>. $\int_a^b F'(x)\,dx = \int_a^b f(x)\,dx = F(b) - F(a)$ is
     one of the most important equations you will encounter in your study of
     mathematics. Be sure you understand how to use it.

2. <u>Notation.</u> Get familiar with $F(x)\Big|_a^b$. It will be seen throughout the course.

3. <u>Proof of the fundamental theorem.</u> The proof depends on the concepts of upper sums, lower sums, telescoping sums, and the mean value theorem. Knowing how to use the theorem is of more immediate importance than understanding the proof.

4. <u>Dummy variable.</u> Changing the variable letter does not change the value of an integral. $\int_a^b f(N)\,dN$ is the same as $\int_a^b f(x)\,dx$.

5. <u>Algebraic manipulations of integrands.</u> Example 7 demonstrates how "hard" integrals may be converted to "easy" integrals. Expansion and division are common simplification techniques. Other methods will be introduced in subsequent chapters.

SOLUTIONS TO EVERY OTHER ODD EXERCISE

1. By the power rule for antidifferentiation, an antiderivative for x^3 is $x^4/4$. Thus $\int_1^3 x^3\,dx = (x^4/4)\Big|_1^3 = (3)^4/4 - (1)^4/4 = (81 - 1)/4 = 20$.

5. $F(x)\Big|_a^b = F(b) - F(a)$; therefore $x^{3/4}\Big|_0^2 = (2)^{3/4} - (0)^{3/4} = 8^{1/4}$.

9. $\int_a^b s^{4/3}\,ds = (3s^{7/3}/7)\Big|_a^b = (3/7)(b^{7/3} - a^{7/3})$. This exercise demonstrates the concept of a dummy variable.

13. The fundamental theorem yields $\int_0^{10}(t^4/100 - t^2)\,dt = (t^5/500 - t^3/3)\Big|_0^{10} = 100000/500 - 1000/3 = -400/3$.

17. From Section 2.5, we get the formula $\int (at + b)^n\,dt = (at + b)^{n+1}/a(n + 1) + C$. Thus, $\int_1^2 [dt/(t + 4)^3] = \int_1^2 (t + 4)^{-3}\,dt = [(t + 4)^{-2}/(-2)]\Big|_1^2 = -1/72 - (-1/50) = 11/1800$.

21. Expand and then perform division on the integrand to get $\int_1^2 [(x^2 + 5)^2/x^4]\,dx = \int_1^2 [(x^4 + 10x^2 + 25)/x^4]\,dx = \int_1^2 (1 + 10x^{-2} + 25x^{-4})\,dx = (x - 10x^{-1} - 25/3x^3)\Big|_1^2 = 3 - 10(-1/2) - (25/3)(-7/8) = 367/24$.

25. The region is that under the graph of $y = x^2$ from $x = 1$ to $x = 2$, so its area is $\int_1^2 x^2 dx$. By the fundamental theorem, $\int_1^2 x^2 dx = (x^3/3)\big|_1^2 = 7/3$. Thus, the area is $7/3$.

29.

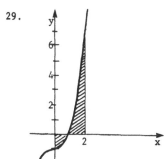

An antiderivative for $x^3 - 1$ is $x^4/4 - x$, so $\int_0^2 (x^3 - 1)dx = (x^4/4 - x)\big|_0^2 = 16/4 - 2 = 2$. The integral represents the signed area between $x^3 - 1$ and the x-axis from $x = 0$ to $x = 2$, as depicted by the shaded region at the left.

33.

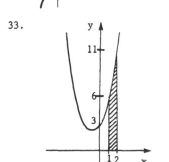

The graph of $x^2 + 2x + 3$ lies above the x-axis, so the area between the function and the x-axis in the interval $[1,2]$ is given by $\int_1^2 (x^2 + 2x + 3)dx = (x^3/3 + x^2 + 3x)\big|_1^2 = 7/3 + 3 + 3 = 25/3$.

37.

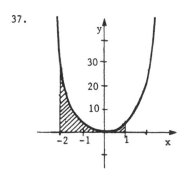

The graph of $x^4 + 3x^2 + 1$ lies above the x-axis, so the area between the function and the x-axis in the interval $[-2,1]$ is given by $\int_{-2}^1 (x^4 + 3x^2 + 1)dx = (x^5/5 + x^3 + x)\big|_{-2}^1 = 33/5 + 9 + 3 = 93/5$.

41. The displacement equals the total distance travelled since $6t^4 + 3t^2 > 0$. Thus, the displacement is $\Delta d = \int_1^{10}(6t^4 + 3t^2)dt = (6t^5/5 + t^3)\big|_1^{10} = 121,000 - 11/5 = (604,989/5)$ units.

45. The distance travelled is the integral of the velocity, which is
$\int_0^{10} 32t\,dt = 16t^2 \big|_0^{10} = 1600$ feet.

49. This exercise completes the proof of the fundamental theorem. Choose
an appropriate partition $(x_0, x_1, x_2, \ldots, x_m)$ and let $\ell_1, \ell_2,$
$\ldots, \ell_m$ be the values of h on the partition intervals. On
(x_{i-1}, x_i), $F'(t) = f(t) \leqslant \ell_i$. By Consequence 1 of the mean value theorem
$[F(x_i) - F(x_{i-1})]/[x_i - x_{i-1}] \leqslant \ell_i$, or $[F(x_i) - F(x_{i-1})] \leqslant \ell_i \Delta x_i$.
Summing for $i = 1$ to m yields a telescoping sum on the left, while the
right-hand side is the integral of h on $[a,b]$. Therefore, $F(b) -$
$F(a) \leqslant \int_a^b h(t)\,dt$.

SECTION QUIZ

1. State the fundamental theorem of calculus.

2. Compute:

   (a) $\int_{-3}^{3} x\,dx$

   (b) $\int_{-2}^{2} x^3\,dx$

   (c) $\int_{-a}^{a} x^n\,dx$, where a is a positive integer and n is an odd, non-
   negative integer.

3. State a theorem using the result of Question 2(c).

4. Is $\int_{-8/9}^{100/17} (x^7 + 6x^6 + 4x^3 + 53)\,dx$ equal to $\int_{-8/9}^{100/17} (A^7 + 6A^6 + 4A^3 + 53)\,dA$?

5. Compute $\int_0^1 x(x+1)(x+2)\,dx$.

6. The bottom of the swimming pool outside your mansion is bounded by
$y = 3 - 2x - x^2$, the lines $x = -1$ and $x = 2$, and the x-axis.
It is to be lined with gold. Since gold is so expensive, you need to
find the exact area. What is it?

ANSWERS TO PREREQUISITE QUIZ

1. (a) $x^2/2 + C$

    (b) $x^5/5 + C$

    (c) $x^{n+1}/(n + 1) + C$

2. (a) $3x^2/2 + C$

    (b) $7x^7 + C$

    (c) $x^{n+1}/2 + C$

    (d) $5x + C$

3. (a) $x^2/2 + 3x + C$

    (b) $x^5/5 + 3x^2/2 + 5x + C$

ANSWERS TO SECTION QUIZ

1. $\int_a^b F'(x)dx = F(b) - F(a)$

2. (a) 0

    (b) 0

    (c) 0

3. If n is an odd integer, then $\int_{-a}^{a} x^n dx = 0$. There are no restrictions
    on a .

4. Yes, only the dummy variable has changed.

5. $\int_0^1 x(x + 1)(x + 2)dx = \int_0^1 (x^3 + 3x^2 + 2x)dx = 9/4$.

6. $\int_{-1}^{2} (3 - 2x - x^2)dx = 3$ square units.

4.5   Definite and Indefinite Integrals

PREREQUISITES

1.   Recall the rules for differentiating sums, products, quotients, and a
     power of a function (Section 1.3).

2.   Recall the properties of summation (Section 4.1).

3.   Recall the fundamental theorem of calculus (Section 4.4).

PREREQUISITE QUIZ

1.   Differentiate the following functions:

   (a) $x^2 + 3x + 1$

   (b) $(x^5 + x)(x^2 + 3x + 1)$

   (c) $(x + 3)/(x^2 + 1)$

   (d) $(x + 10)^3/x$

2.   Fill in the blank.

   (a) $\displaystyle\sum_{i=1}^{25} (i^2 + i + 1) = \sum_{i=1}^{25} (i + 1) +$ _____ .

   (b) $\displaystyle\sum_{i=1}^{7} (6/i) =$ _____ $\times \displaystyle\sum_{i=1}^{7} (1/i)$.

   (c) $\displaystyle\sum_{k=3}^{10} k = \sum_{k=3}^{5} k +$ _____ .

3.   Compute $\displaystyle\sum_{j=0}^{15} 4$.

4.   What is the relationship between the integral and the derivative?

GOALS

1.   Be able to state the alternative version of the fundamental theorem of
     calculus.

2.   Be able to use the properties of integration to compute integrals.

3.   Be able to check any integration formula by differentiating.

STUDY HINTS

1. Definite vs. indefinite integrals. The box on p. 232 relates the
   definite integral, $\int_a^b f(x)dx$, on the right-hand side to the indefinite
   integral, $\int f(x)dx$, on the left. The box simply shows another way to
   express the fundamental theorem of calculus. You should recognize that
   definite integrals specify their endpoints and possess a unique value.
   Remember that indefinite integrals always have an additive constant.

2. Indefinite integral test. The value of this test cannot be emphasized
   enough. Since differentiation is much easier than integration, it is a
   good habit to check your integrals by differentiating.

3. Properties of integration. You should become familiar with the statements made in the box on p. 234. As with the properties of summation,
   the properties of integration will be memorized with use. Notice how
   Properties 1 and 2 are similar to the differentiation rules with the
   same name.

4. Products of integrals. In general, the integral of a product is not
   the product of each multiplicand's integral, i.e., $\int_a^b f(x)g(x)dx \neq$
   $\int_a^b f(x)dx \cdot \int_a^b g(x)dx$. The technique for integrating products will be
   shown in Section 7.4 .

5. Proof of the properties. Your instructor will probably not hold you
   responsible for proofs such as those in Examples 3 and 4, but ask to
   be sure.

6. "Wrong-way" integrals. There is no need to memorize these formulas.
   They can easily be recalled by using the fundamental theorem of calculus.
   Notice that $\int_a^a f(x)dx = F(a) - F(a) = 0$.

7. Alternative version of the fundamental theorem. This version is not as
    important as $\int_a^b f(x)dx = F(b) - F(a)$ for this course. Note that dif-
    ferentiation is performed with respect to the upper endpoint. Use the
    concept of "wrong-way" integrals to differentiate with respect to the
    lower endpoint. For differentiation with respect to the lower limit,
    you can write $\int_x^b f(s)ds = -\int_b^x f(s)ds$.

8. Geometric interpretation of the fundamental theorem. Fig. 4.5.3 should
    help explain why the alternative version of the fundamental theorem is
    true, but the most important immediate goal is to understand the state-
    ment in the box on p. 237 .

9. Generalization of the fundamental theorem. Use the chain rule and the
    fundamental theorem of calculus to derive $(d/dt)\int_a^{g(t)} f(s)ds = f(g(t)) \times$
    $g'(t)$. Notice that if $g(t) = t$, the original version results.
    See Exercise 43.

SOLUTIONS TO EVERY OTHER ODD EXERCISE

1. Use the indefinite integral test. $(d/dx)(x^5 + C) = 5x^4$, which is the
    integrand, so the formula is correct.

5. (a) Use the indefinite integral test with the quotient rule.
    $(d/dt)[t^3/(1 + t^3) + C] = [3t^2(1 + t^2) - t^3(3t^2)]/(1 + t^3)^2 =$
    $3t^2/(1 + t^3)^2$, which is the integrand, so the formula is correct.

    (b) Apply the fundamental theorem of calculus to get $\int_0^1 [3t^2/(1 + t^3)^2]dt =$
    $[t^3/(1 + t^3)]\Big|_0^1 = 1/2$.

9. By the fundamental theorem of calculus, $\int_{-2}^3 (x^4 + 5x^2 + 2x + 1)dx =$
    $(x^5/5 + 5x^3/3 + x^2 + x)\Big|_{-2}^3 = (243 + 32)/5 + 5(27 + 8)/3 + (9 - 4) +$
    $(3 + 2) = 370/3$.

13. Divide first to get $\int_1^2 [(x^2 + 2x + 2)/x^4] dx = \int_1^2 (x^{-2} + 2x^{-3} + 2x^{-4}) dx = (-x^{-1} - x^{-2} - 2x^{-3}/3) \Big|_1^2 = (-1/2 - 1/4 - 1/12) - (-1 - 1 - 2/3) = 11/6$.

17. Guess that an antiderivative is $(1 + 2t)^6 + C$. The indefinite integral test yields $(d/dx)[(1 + 2t)^6 + C] = 2(6)(1 + 2t)^5$, so the antiderivative must be $(1/12)(1 + 2t)^6 + C$. By the fundamental theorem, $\int_1^2 (1 + 2t)^5 dt = (1/12)(1 + 2t)^6 \Big|_1^2 = (1/12)(5^6 - 3^6) = (1/12)(15625 - 729) = 3724/3$.

21. Using Property 3 of integration, we get $\int_0^2 f(x) dx = \int_0^1 f(x) dx + \int_1^2 f(x) dx = 3 + 4 = 7$.

25. This is a wrong-way integral. $\int_3^2 x dx = (x^2/2) \Big|_3^2 = (4 - 9)/2 = -5/2$.

29. The left-hand side is $(d/dx)\int_a^x (s^3 - 1) ds = (d/dx)[(s^4/4 - s) \Big|_a^x] = (d/dx)(x^4/4 - x - a^4/4 + a) = x^3 - 1 = f(x)$ because a is constant.

33. According to the alternative version of the fundamental theorem, $(d/dt)\int_a^t f(x) dx = f(t)$. When $f(x) = 3/(x^4 + x^3 + 1)^6$, the answer is $3/(t^4 + t^3 + 1)^6$.

37. Differentiating the distance function, $\int_a^t v(s) ds$, with respect to time yields the velocity function.

41. (a) Let $F(s)$ be an antiderivative for $f(s)$, then $F_1(t) = F(s) \Big|_{a_1}^t$ and $F_2(t) = F(s) \Big|_{a_2}^t$. $F_1(t) - F_2(t) = [F(t) - F(a_1)] - [F(t) - F(a_2)] = F(a_2) - F(a_1)$. Since a_1 and a_2 are constants, $F(a_2) - F(a_1)$ must be a constant.

    (b) From part (a), $F_2(t) - F_1(t) = F(a_1) - F(a_2)$. By the fundamental theorem of calculus, this is $\int_{a_2}^{a_1} f(s) ds$.

45. The generalized chain rule was derived in Exercise 43. It states that $(d/dx)\int_a^{g(x)} f(t) dt = f(g(x)) \cdot g'(x)$. Here, $g(x) = x^2$, $a = 1$, and $f(t) = 1/t$. Therefore, $F'(x) = (1/x^2) 2x = 2/x$.

49.   Let $I = \int_a^b f(t)dt + \int_b^c f(t)dt$.   We will show that every number less than

I is a lower sum for f on [a,c] and every number greater than I

is an upper sum for f on [a,c] ; by the definition of the integral,

f will be integrable on [a,c] and its integral will be I .

So let $S < I$. To show that S is a lower sum, we begin by writing

$S = S_1 + S_2$, where $S_1 < \int_a^b f(t)dt.$ and $S_2 < \int_b^c f(t)dt$.   (See the hint on

p. A.29.) This means that S_1 and S_2 are lower sums for f on [a,b] and

[b,c] , respectively. Thus there is a step function g_1 on [a,b] with

$g_1(t) \leqslant f(t)$ for all t in (a,b) , and $\int_a^b g_1(t)dt = S_1$, and there is a

step function g_2 on [b,c] with $g_2(t) \leqslant f(t)$ for all t in (b,c) and

$\int_b^c g_2(t) = S_2$. Put together g_1 and g_2 to obtain a function g on [a,c] by

the definition:

$$g(t) = \begin{cases} g_1(t) & a \leqslant t < b \\ f(b) & t = b \\ g_2(t) & b < t \leqslant c \end{cases}.$$

The function g is a step function on [a,c] , with $g(t) \leqslant f(t)$ for

all t in (a,c) . The sum which represents the integral for g on

[a,c] is the sum of the sums representing the integrals of g_1 and g_2 ,

so we have

$$\int_a^c g(t)dt = \int_a^b g_1(t)dt + \int_b^c g_2(t)dt = S_1 + S_2 = S$$

and S is a lower sum. Similarly, any number greater than I is an

upper sum, so I is the integral of f on [a,b] .

Remark. Using the definition of "wrong-way integrals," one may easily
verify that Property 3 holds no matter how a , b , and c are ordered.

53. Since f is continuous, we can take the limit of the expression in

Exercise 51. Using the definition of the derivative, we have

$F'(t) = \lim_{h \to 0} \{[F(t + h) - F(t)]/h\} = \lim_{h \to 0} [(1/h) \int_t^{t+h} f(s)ds]$. From Exer-

cise 52, the limit on the right equals f(c) for $t \leqslant c \leqslant t + h$,

but as $h \to 0$, $c \to t$. Therefore, the limit on the right is f(t)

as we wished to show.

SECTION QUIZ

1. Check the following integration formulas:

(a) $\int (5x + 3)(x + 1)^3 dx = (x + 1/2)(x + 1)^4 + C$.

(b) $\int (3 + 4x)^7 dx = (3 + 4x)^8/8 + C$.

(c) $\int (x^4 + 2x^2)dx = 4(x^3 + x) + C$

2. Compute $\int_0^2 2x dx$, and $\left[\int_0^2 2dx\right] \cdot \left[\int_0^2 x dx\right]$. What do your results imply

about the product of integrals?

3. Simplify the following expression to a single integral: $\int_0^3 f(x)dx +$

$\int_3^7 f(x)dx - \int_7^0 f(x)dx$.

4. Let $G(t) = \int_{-t}^{t^2} (x + 3)^3 dx$. What is $G'(1)$?

5. A man buys some beachfront property for his alligator farm. If he puts

a fence at the x-axis, the shoreline is given by $f(x) = x^3/3 - x + 1$

at high tide. He wants to know how much sand his alligators have to

crawl around in.

(a) Find the area for $-1 \leqslant x \leqslant 1$.

(b) He decides to buy more land from x = 1 to x = 2 . Use Property

3 of integration to compute the area for $-1 \leqslant x \leqslant 2$.

(c) At low tide, the shoreline has the shape 2f(x) . Find the area

on [-1,2] .

(d) How fast is the area changing (with respect to x) at x = 1 ?

ANSWERS TO PREREQUISITE QUIZ

1.   (a) $2x + 3$

 (b) $(5x^4 + 1)(x^2 + 3x + 1) + (x^5 + x)(2x + 3)$

 (c) $(-x^2 - 6x + 1)/(x^2 + 1)^2$

 (d) $(2x - 10)(x + 10)^2/x^2$

2.   (a) $\sum\limits_{i=1}^{25} i^2$

 (b)   6

 (c) $\sum\limits_{k=6}^{10} k$

3.   64

4. $\int_a^b f(x)\,dx = F(x)\Big|_a^b$ where $F'(x) = f(x)$.

ANSWERS TO SECTION QUIZ

1.   (a)   The formula is correct.

 (b)   The denominator should be 32 rather than 8 .

 (c)   The right-hand side is the derivative when $C = 0$, not the integral.

2. $\int_0^2 2x\,dx = 4$, $(\int_0^2 2\,dx)(\int_0^2 x\,dx) = (4)(2) = 8$. In general, the product of
 two integrals does not equal the integral of the product of the integrands,
 i.e., $\int_a^b f(x)g(x)\,dx \neq [\int_a^b f(x)\,dx] \cdot [\int_a^b g(x)\,dx]$.

3. $2\int_0^7 f(x)\,dx$

4. $G'(t) = (t^2 + 3)^3(2t) + (-t + 3)^3$, so $G'(1) = 136$.

5.   (a)   2

 (b) $\int_{-1}^2 f(x)\,dx = \int_{-1}^1 f(x)\,dx + \int_1^2 f(x)\,dx = 2 + 3/4 = 11/4$.

 (c) $\int_{-1}^2 2f(x)\,dx = 2(11/4) = 11/2$.

 (d)   1/3

4.6   Applications of the Integral

PREREQUISITES

1.   Recall how to calculate the area between a positive function and the
     x-axis (Sections 4.2 and 4.4).

PREREQUISITE QUIZ

1.   Use integration to compute the area under $y = 3$ on $[-1,2]$.

2.   Find the area under the curve $y = (x^3 + 1)/x^2$ for $1 \leqslant x \leqslant 2$.

3.   Calculate the area of the shaded region,
     which is bounded by $y = -x^2 + 3x + 4$.

GOALS

1.   Be able to compute the area of a region between two curves.

2.   Be able to compute the total change in a quantity from its rate of change.

STUDY HINTS

1.   Area versus signed area. Remember that the integral gives the signed
     area; therefore, the area of regions where the function is negative is
     the negative of the integral for the same region, i.e., $-\int_a^b f(x)dx$
     wherever $f(x) < 0$.

2. Area between curves. You should memorize the formula. Always subtract
    the lower curve from the higher curve. The formula works whether or not
    the curves lie above the x-axis. Note that if $f(x) = 0$, we get the
    formula for the area under the positive function, $g(x)$.

3. Infinitesimal approach. Try to understand this approach for deriving
    the area formula. Learning this method will help you derive formulas
    which will be introduced in future chapters as well as reduce the risk
    of memory lapses during exams.

4. Intersecting curves. It is a good idea to sketch all regions in area
    problems to check for points of intersection. These points must be
    known to determine the endpoints of integration. Again, in the area
    formula, subtract the lower curve from the higher one on each interval.

5. Multi-functional boundaries. If an area is bounded by several functions,
    it must be divided into subregions which are bounded by only two functions.
    Then, apply the additive property of areas.

6. Symmetry. Noting some symmetry can simplify a problem. For example, the
    area under $y = x^2$ on $[-1,1]$ is $2\int_0^1 x^2 dx$.

7. Integrating rates. Know that integrating rates of change gives the total
    change. Your instructor may choose to emphasize some examples over others,
    but only the terminology differs. For example, the rate, $r(x)$, may be
    called velocity, marginal revenue, current, productivity, or many other
    names depending on the application.

SOLUTIONS TO EVERY OTHER ODD EXERCISE

1.

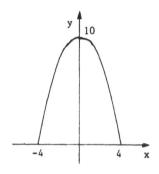

Place the parabolic arch as shown at the left. The curve has the equation $y = ax^2 + b$. When $x = 0$, we get $b = 10$. At the x-axis, we have $0 = 16a + 10$, so $a = -5/8$. Thus, the enclosed area is $\int_{-4}^{4}(-5x^2/8 + 10)dx = (-5x^3/24 + 10x)\big|_{-4}^{4} = -80/3 + 80 = 160/3$.

5. $x^2/4 - 1 = 1 - x^2/4$ implies $x = \pm 2$ are the intersection points of the two curves. The area is $\int_{-2}^{2}[(1 - x^2/4) - (x^2/4 - 1)]dx = \int_{-2}^{2}(2 - x^2/2)dx = (2x - x^3/6)\big|_{-2}^{2} = 8 - 8/3 = 16/3$.

9. On the interval $[0,1]$, $\sqrt{x} \geqslant x$, so the area is $\int_{0}^{1}(x^{1/2} - x)dx = (2x^{3/2}/3 - x^2/2)\big|_{0}^{1} = 1/6$.

13. $3x^2 \leqslant 3/4$ in $[-1/2,1/2]$ while $x^4 + 2 \geqslant 2$ on the same interval, so the area is $\int_{-1/2}^{1/2}(x^4 + 2 - 3x^2)dx = (x^5/5 + 2x - x^3)\big|_{-1/2}^{1/2} = (1/160 + 1 - 1/18) - (-1/160 - 1 + 1/18) = 141/80$.

17.

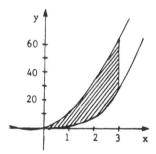

The area of the region sketched at the left is $\int_{0}^{3}(5x^2 + 6x - x^3)dx = (5x^3/3 + 3x^2 - x^4/4)\big|_{0}^{3} = 45 + 27 - 81/4 = 207/4$.

21. Use Method 2 of Example 8. $y^2 - 3 = 2y$ gives the limits of integration as 3 and -1 . Since $2y \geqslant y^2 - 3$ on $[-1,3]$, the area is $\int_{-1}^{3}[2y - (y^2 - 3)]dy = (y^2 - y^3/3 + 3y)\big|_{-1}^{3} = 8 - 28/3 + 12 = 32/3$.

25. Use the formula $\Delta Q = Q(b) - Q(a) = \int_a^b r(t)dt$. Here, $r(t) = 300t^2$,

a = 0 , and b = 5 . Therefore, the number of liters released is

$\int_0^5 300t^2 dt = 100t^3 \big|_0^5 = 100(125) = 12,500$ liters.

29. (a)

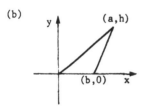

The equation of the line segment from (0,0)

to (a,h) is y = (h/a)x . For the seg-

ment from (a,h) to (b,0) , it is

y = [-h/(b - a)](x - b) . The area is

$\int_0^a (h/a)x\,dx + \int_a^b [-h/(b - a)](x - b)dx =$

$(h/a)x^2/2 \big|_0^a + [-h/(b - a)][x^2/2 - bx]\big|_a^b = ha/2 + (-h)[(b - a)/2 - b] =$

hb/2 . Geometrically, the height is h and the base is b ; there-

fore, the area is (1/2)(base)(height) = bh/2 . Thus, both methods

give the same answer.

(b)

Again, the equations are y = (h/a)x and

y = [-h/(b - a)](x - b) . The area is

$\int_0^b (h/a)x\,dx + \int_b^a [(h/a)x - ((-h)/(b - a)) \times$

$(x - b)]\,dx = (h/a)(x^2/2)\big|_0^b + (h/a)(x^2/2)\big|_b^a +$

$[h/(b - a)](x^2/2 - bx)\big|_b^a = hb^2/2a + ha/2 -$

$hb^2/2a + (-h)[(a + b)/2 - b] = bh/2$. Geometrically, the height is

h and the base is b , so the area is bh/2 .

33. The equation of a parabolic curve is $y = cx^2$. Put the corner of the para-

bola at the origin, then the center of the roof is located at (12,40) .

Therefore, c must be 3/400 . The area of the wall with the exhaust

fans is $(20)(80) + 2\int_0^{40}(3x^2/400)dx = 1600 + (3/200)(x^3/3)\big|_0^{40} = 1600 +$

320 = 1920 , so the volume is (1920)(180) = 345,600 cubic feet. The

four fans can move 4(5500) = 22000 cubic feet. The elapsed time is

345,600/22,000 = 15.71 minutes or 15 minutes, 43 seconds.

37. Let the horizontal line be drawn at $y = 1/c^2$, then the area of the

upper region is $\int_1^c (1/x^2 - 1/c^2)dx = (-x^{-1} - x/c^2)\vert_1^c = -1/c - 1/c + 1 +$

$1/c^2 = 3/8$. Algebraic manipulations yield $5c^2/8 - 2c + 1 = 0$. The

solution is $(8 + 2\sqrt{6})/5$ because $1 < c < 4$. $1/c = 5/(8 + 2\sqrt{6}) =$

$(4 - \sqrt{6})/4$ by rationalizing the denominator. Finally, we square to

get $1/c^2 = y = (11 - 4\sqrt{6})/8$.

SECTION QUIZ

1.

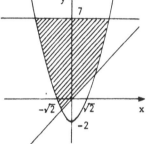

Find the area of the shaded region which is

bounded by the curves $y = 7$, $y = x^2 - 2$,

and $y = x$.

2. Find the area between $y = x^2 - 4x$ and $y = -3$ on $[-1,4]$.

3. The Hi-Ho Mining Company produces coal. The head dwarf at the company

    determines that at a depth of x meters, they can get $6 + 2x$ tons of

    coal.

    (a) How much can be mined in the first 20 meters?

    (b) How much can be mined at a depth of exactly 50 meters?

4. Mr. and Mrs. Chip have just been blessed with a handsome cookie monster

    child. Since consumption of cookies causes expansion of cookie monsters'

    stomachs, the child can eat $(x^2/2 + 10)$ thousand cookies per year in

    the xth year of life. How many cookies do Mr. and Mrs. Chip need to

    buy for their child between the ages of 7 and 14 ?

ANSWERS TO PREREQUISITE QUIZ

1. $\int_{-1}^{2} 3dx = 9$

2.   2

3. $\int_{-1}^{4} (-x^2 + 3x + 4)dx = 125/6$

ANSWERS TO SECTION QUIZ

1.   63/2

2.   28/3

3.   (a)   520 tons

     (b)   106 tons

4. $\int_{7}^{14} (x^2/2 + 10)dx = (2821/6)$ thousand cookies

4.R   Review Exercises for Chapter 4

SOLUTIONS TO EVERY OTHER ODD EXERCISE

1. $\sum_{i=1}^{4} i^2 = (1)^2 + (2)^2 + (3)^2 + (4)^2 = 1 + 4 + 9 + 16 = 30$.

5. Using the properties of summation, $\sum_{i=1}^{500} (3i + 7) = 3 \sum_{i=1}^{500} i + \sum_{i=1}^{500} 7$. Then,

    applying the formula $\sum_{i=1}^{n} i = n(n + 1)/2$ gives $3(500)(501)/2 + 7(500) =$

    $375,750 + 3500 = 379,250$.

9. This is a step function. Using the partition $(0,1/5,1/4,1/3,1/2,1)$,

    we have $\int_0^1 f(x)dx = 1(1/5 - 0) + 2(1/4 - 1/5) + 3(1/3 - 1/4) + 4(1/2 -$

    $1/3) + 5(1 - 1/2) = 1/5 + 1/10 + 1/4 + 2/3 + 5/2 = 223/60$.

13. Using the fundamental theorem of calculus, $\int_3^5 (-2x^3 + x^2)dx = (-x^4/2 +$

    $x^3/3)\big|_3^5 = (-625 + 81)/2 + (125 - 27)/3 = -272 + 98/3 = -718/3$.

17. On the interval $[0,1]$, $x^3 + x^2 \geqslant 0$, so the area under the curve is

    $\int_0^1 (x^3 + x^2)dx = (x^4/4 + x^3/3)\big|_0^1 = 1/4 + 1/3 = 7/12$.

21. (a) If intervals of equal length are used, then at least 10 intervals

         are required to satisfy the condition that the upper and lower sums

         are within 0.2 of one another. In this case, the upper sum is

         $\sum_{i=0}^{9} [4/(1 + (i/10)^2)] (1/10) = 3.2399$.   The lower sum is

         $\sum_{i=1}^{10} [4/(1 + (i/10)^2)] (1/10) = 3.0399$.

    (b) The average of 3.2399 and 3.0399 is 3.1399 , so we guess that

         the exact value of the integral is approximately 3.1399 . (In fact,

         the actual value of the integral is π .)

25. Using calculus, the area is $\int_{a_1}^{a_2}(mx + b)dx = (mx^2/2 + bx)\Big|_{a_1}^{a_2} =$

    $m(a_2^2 - a_1^2)/2 + b(a_2 - a_1) = [(a_2 - a_1)/2](ma_1 + b + ma_2 + b)$.

    Alternatively, the area under the curve is a trapezoid. Plane

    geometry tells us that $A = (h/2)(b_1 + b_2) = [(a_2 - a_1)/2][(ma_1 + b) +$

    $(ma_2 + b)]$, which is exactly the same. (b_1 is $f(a_1)$ and b_2 is

    $f(a_2)$.)

29. (a) By the indefinite integral test on p. 233, we must differentiate

    the right-hand side and see if we get the integrand $x^2/(x^3 + 6)^2$.

    By the quotient rule, the derivative of the right-hand side is

    $(1/12)[(3x^2)(x^3 + 6) - (x^3 + 2)(3x^2)]/(x^3 + 6)^2 =$

    $(1/12)(12x^2)/(x^3 + 6)^2 = x^2/(x^3 + 6)^2$; therefore, the formula

    is verified by the indefinite integral test.

    (b) The area is $\int_0^2[x^2/(x^3 + 6)^2]dx = (1/12)[(x^3 + 2)/(x^3 + 6)]\Big|_0^2 =$

    $(1/12)(10/14 - 2/6) = (1/12)(8/21) = 2/63$.

33. This is best done by integrating in y . $y^2 - 6 = y$ implies that the

    limits of integration are -2 and 3 . On $[-2,3]$, $y \geqslant y^2 - 6$, so

    the area is $\int_{-2}^3[y - (y^2 - 6)]dy = (y^2/2 - y^3/3 + 6y)\Big|_{-2}^3 = 5/2 - 35/3 +$

    $30 = 125/6$.

37. In each case, let $L(t)$ be the amount of leakage. Then as long as the

    tank does not empty for more than an instant, the volume of water at

    the end of 3 minutes is given by $1 + \int_0^3(3t^2 - 2t + 3 - L)dt$. If the

    tank does empty, then the volume is given by $\int_x^3(3t^2 - 2t - 3 - L)dt$,

    where x is the time the tank begins to fill again.

    (a) Here, the rate of volume increase is $3t^2 - 2t + 1$, which is

    always positive, so $V(3) = 1 + \int_0^3(3t^2 - 2t + 1)dt = 1 + (t^3 - t^2 +$

    $t)\Big|_0^3 = 1 + 27 - 9 + 3 = 22$ liters.

37. (b) The rate of volume increase is $3t^2 - 2t - 1$, which is $\leqslant 0$ for

   $-1/3 \leqslant t \leqslant 1$. However, $1 + \int_0^x (3t^2 - 2t - 1)dt = 1 + x^3 - x^2 -$

   $x = x^2(x - 1) + (-1)(x - 1) = (x^2 - 1)(x - 1)$. This shows that

   volume is 0 only at $t = 1$ minute, so $V(3) = 1 + (t^3 - t^2 -$

   $t)|_0^3 = 1 + 27 - 9 - 3 = 16$ liters.

 (c) The rate of volume increase is $3t^2 - 2t - 5$, which is $\leqslant 0$ for

   $-1 \leqslant t \leqslant 5/3$. If we use $V(x) = 1 + \int_0^x (3t^2 - 2t - 5)dt$, then

   $V(1) = -4$, so the tank must be empty for more than an instant.

   Therefore, $V(3) = \int_{5/3}^3 (3t^2 - 2t - 5)dt = (t^3 - t^2 - 5t)|_{5/3}^3 =$

   $3 - (-175/27) = (256/27)$ liters.

41. The distance travelled is $\int_a^b |v(t)|dt$. Velocity is 0 at $t = -1$

 and 5 , which is where velocity changes direction. Thus, the distance

 travelled is $-\int_0^5 (t^2 - 4t - 5)dt + \int_5^6 (t^2 - 4t - 5)dt = -(t^3/3 - 2t^2 -$

 $5t)|_0^5 + (t^3/3 - 2t^2 - 5t)|_5^6 = 100/3 + 10/3 = 110/3$.

45. (a) $C(t) = \int (1000t - 7000)dt = 500t^2 - 7000t + K$. $C(0) = 40,000$

   implies $K = 40,000$. We want to solve $C(t) = 20,000$, so

   $20,000 = 500t^2 - 7000t + 40,000$ implies $0 = 500(t^2 - 14t + 40) =$

   $500(t - 4)(t - 10)$. Due to the restriction that $0 \leqslant t \leqslant 6$, the

   only solution is $t = 4$. Thus after 4 days, the concentration

   has dropped to half its original value, so the inspector should be

   sent on the fifth day.

 (b) The total change in concentration from the fourth to the sixth day

   is $\int_4^6 1000(t - 7)dt = 1000\int_4^6 (t - 7)dt = 1000(t^2/2 - 7t)|_4^6 =$

   $1000(10 - 14) = -4000$. Therefore, the concentration drops by

   4000 bacteria per cubic centimeter.

49. By the alternative version of the fundamental theorem of calculus,

$(d/dx)\int_0^x [s^2/(1 + s^3)] ds = x^2/(1 + x^3)$.

53. Apply the generalized version of the fundamental theorem, which states

$(d/dt)\int_a^{g(t)} f(y)dy = f(g(t))\cdot g'(t)$. Here $g(t) = t^3 + 2$, $a = 0$,

and $f(y) = 1/(y^2 + 1)$. Thus, the derivative is $3t^2/[(t^3 + 2)^2 + 1]$.

TEST FOR CHAPTER 4

1.   True or false.

(a) $\sum_{i=0}^{10} 3 = 33$.

(b)   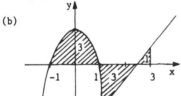   The areas of the shaded humps are 3 ,

3 , and 1 , respectively; therefore,

$\int_{-1}^3 f(x)dx = 7$.

(c)   If $F(x) = \int_7^x (dt/t)$, then $F'(x) = (1/t)\big|_7^x$.

(d) $\sum_{i=1}^{10} 2^i + \sum_{k=1}^{19} (1/k)$ has no value because the dummy indices are

different.

(e) $\int x^2(x^3 + 2)^4 dx = (x^3 + 2)^5/15 + C$.

2.   Circle the correct conclusion.

(a)   The fundamental theorem of calculus states that

(i) $\int_a^b f(x)dx = f'(b) - f'(a)$.

(ii) $\int_a^b f'(x)dx = f(b) - f(a)$.

(iii) $\int_a^b f(x)dx = F'(x)\big|_a^b$ where $F'(x) = f(x)$.

2. (b) The fundamental theorem of calculus states that

(i) $(d/dx)\int_a^x f(t)dt = f(t)$.

(ii) $(d/dx)\int_a^x f(t)dt = f(x)$.

(iii) $(d/dx)\int_x^a f(t)dt = f(x)$.

(iv) $(d/dx)\int_a^x f(t)dt = F(x)$ where $F'(x) = f(x)$.

3. Let $f(x) = (x^3 + 6)/(x^2 - x + 3)$; what is $\int_{-1}^1 f'(u)du$?

4. Compute:

(a) $\int_{60}^{50}\left(\sum_{i=1}^{19} i\right)dx$

(b) $(d/dx)\int_{17}^{19}(t^5 + 4t^4 + 2t^2 + 4)dt$

(c) $\sum_{i=5}^{48} [i/(i - 1) - (i - 2)/(i - 3)]$. Leave your answer as a sum of

fractions.

(d) $\int_0^1 [(x^2 + x - 6)/(x + 3)]dx$

5. Show that $\int_3^5 [(x + 5)/(x + 2)]dx \leqslant 16/5$.

6. There is only one region bounded by the ellipse $x^2/9 + y^2/4 = 1$ and

$x = |1|$ which contains the origin. Express the area of this region

as an integral, but don't evaluate it.

7. Find the area of the region bounded by $y = x$ for $x \geqslant 0$, $y = 5x$

for $x \leqslant 0$, and $6 - x^2 = y$ for all x .

8. (a) Differentiate $(2x + 3)^3(3x + 1)^2$.

(b) Compute $\int_0^1 3(2x + 3)^2(3x + 1)(5x + 4)dx$.

9. Evaluate the following integrals:

(a) $\int [(t^4 - 3t^2 + 1)/t^2]dt$

(b) $\int_5^{10} f(x)dx + \int_{10}^{20} f(t)dt + \int_{20}^5 f(s)ds$, where $f(x) = x^5 + 3 - 1/x^7$.

(c) $\int_1^2 (t + 1)(t - 1)dt$

10. Breakout Bob has just escaped into the West. The sheriff immediately
    mounts his horse and starts after Bob. The horse runs at $(20 - t^2)$
    kilometers/hour. After 240 minutes, the sheriff lassos Bob off his
    horse. How far did the sheriff chase after Bob?

ANSWERS TO TEST FOR CHAPTER 4

1. (a) True

    (b) False; it is 1 .

    (c) False; $F'(x) = 1/x$, not $1/x - 1/7$.

    (d) False; the name given to an index does not affect the value of
         a sum.

    (e) True

2. (a) (ii)

    (b) (ii)

3. 4/3

4. (a) -1900

    (b) 0

    (c) 48/47 + 47/46 - 4/3 - 3/2

    (d) -3/2

5. $(x + 5)/(x + 2) \leqslant 8/5$ on [4,5] . Use Property 5 of integration.

6. $4\int_{-1}^{1} \sqrt{1 - x^2/9} \ dx = 8\int_{0}^{1} \sqrt{1 - x^2/9} \ dx$.

7. 184/3

8. (a) $6(2x + 3)^2 (3x + 1)(5x + 4)$

    (b) 1973/2

9.   (a) $t^3/3 - 3t - 1/t + C$

     (b)   0

     (c)   4/3

10. (176/3) kilometers

CHAPTER 5

TRIGONOMETRIC FUNCTIONS

5.1 Polar Coordinates and Trigonometry

PREREQUISITES

1. There are no prerequisites for this section other than basic high
   school geometry.

PREREQUISITE QUIZ

1. Use Quiz C on p. 14 to evaluate your preparation for this section. If
   you pass Quiz C, the material of this section may be reviewed quickly.
   If you have difficulties with Quiz C, you will need to study this section
   thoroughly.

GOALS

1. Be able to plot in polar coordinates.

2. Be able to write down the definitions of the trigonometric functions
   and to be able to find commonly used values.

3. Be able to make sketches of the graphs of the trigonometric functions.

STUDY HINTS

1. Radians. Calculus uses radians, not degrees! Be sure your calculator
   is in the right mode. Remember that 2π radians is $360°$ in order to
   make conversions.

2. __Equivalences.__ You should be able to convert back and forth from radians
    and degrees quickly in your head for the following angles: $0°$, $30°$, $45°$,
    $60°$, $90°$, $180°$, $270°$, $360°$. Use proportional reasoning: $90° = 360°/4$
    is equivalent to $2\pi/4 = \pi/2$.

3. __Negative polar coordinates.__ Remember that a negative angle represents
    clockwise motion and that a negative radius means reflection through the
    origin or a rotation of π radians, i.e., $(-r,\theta) = (r,\theta + \pi)$.

4. __Definitions.__ Using the figure at the left, you should know the following

definitions: $x/r = \cos\theta$, $y/r = \sin\theta$, $y/x =$
$\tan\theta$, $\cot\theta = 1/\tan\theta$, $\sec\theta = 1/\cos\theta$, and
$\csc\theta = 1/\sin\theta$. Note that if $r = 1$, $x = \cos\theta$
and $y = \sin\theta$. Also, no cofunction is the reciprocal
of another cofunction. For example, $\csc\theta$ and $\cos\theta$ are __not__ reciprocals. This may help you remember that $\sec\theta$ and $\cos\theta$ __are__ reciprocals.

5. __Commonly used values.__ Reproducing Fig. 5.1.17 and Fig. 5.1.18 will help
    you remember the most commonly used trigonometric values. If you have
    troubles remembering whether $\cos(\pi/6)$ is $1/2$ or $\sqrt{3}/2$, recall that
    $\cos\theta$ corresponds to x in the xy-plane. The cosine function decreases
    as θ increases on $[0,\pi/2]$. Thus, we expect $\cos(\pi/6) = \sqrt{3}/2$ and
    $\cos(\pi/3) = 1/2$. Similar analysis may be used for sine.

6. __Calculuator errors.__ Inaccuracies may occur if θ is very large. For
    example, does $\cos(201\pi/2) = 0$ on your calculator?

7. __Trigonometric identities.__ For the purposes of this course, you will find
    that $\cos^2\theta + \sin^2\theta = 1$ is very useful. Formulas (2), (3), and (4) will
    help you to derive other identities, but it is not an essential part of
    the course. Formulas (5) and (6) are especially important for integration

7. Trigonometric identities (continued).

    techniques which will be presented in Chapter 10. Other useful trigono-
    metric identities may be found on the inside front cover of the text.

8. Period and frequency. A period is the length of a repeatable unit of a
    trigonometric graph. Frequency is the reciprocal of period.

9. Graphing. Let TRIG be an arbitrary trigonometric function and let A
    and B be constants. The graph of $y = A$ TRIG Bx can be obtained by
    drawing $y =$ TRIG x and relabelling the x- and y-axes. All of the
    y-values should be multiplied by A and the x-values should be divided
    by B . The effect is stretching and compressing.

SOLUTIONS TO EVERY OTHER ODD EXERCISE

1. The problem is analogous to Example 1. We use the formulas $C_\theta = r\theta$ and
    $A_\theta = r^2\theta/2$. Remember that θ must be expressed in radians. We convert
    $22°$ to $22\times(\pi/180) = 0.3839$ radians. Thus, $C_\theta = r\theta = 10(0.3839) =$
    3.839 m , and $A_\theta = (10)^2(0.3839)/2 = 19.195$ m^2 .

5. Since 2π radians $= 360°$, the conversion factor is $2\pi/360 = \pi/180$.
    Thus, $29° = 29(\pi/180) = 0.5061$ rad ; $54° = 54(\pi/180) = 0.9425$ rad ;
    $255° = 255(\pi/180) = 4.4506$ rad; $130° = 130(\pi/180) = 2.2689$ rad ; $320° =$
    $320(\pi/180) = 5.5851$ rad .

9.

In the notation (r,θ) , r refers to the
distance from the origin. If $r < 0$,
rotate the point by π radians or reflect
it across the origin. θ refers to the
angle measured counterclockwise from the

positive x-axis if $\theta > 0$; clockwise if $\theta < 0$.

13.

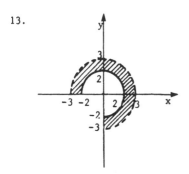

The region lies between circles of radii
2 and 3 , centered at the origin. The
circle of radius 2 is also included. In
addition, the region is further restricted
in that it must lie between angles of
$-\pi/2$ and π , inclusive.

17. To make the conversion from cartesian coordinates to polar coordinates,
use $r = \sqrt{x^2 + y^2}$, $\cos \theta = x/r$, and $\sin \theta = y/r$. θ can be found
by using a trigonometric table or a calculator.

(a) $r = \sqrt{1^2 + 0^2} = 1$; $\cos \theta = 1$ implies $\theta = 0$, so the polar
coordinates are (1,0) .

(b) $r = \sqrt{3^2 + 4^2} = 5$; $\cos \theta = 3/5$ implies $\theta = \pm 0.927$, but $\sin \theta =$
4/5 implies $\theta = 0.927$, so the polar coordinates are (5,0.927) .

(c) $r = \sqrt{3 + 1} = 2$; $\cos \theta = \sqrt{3}/2$ implies $\theta = \pm\pi/6$, but $\sin \theta =$
1/2 implies $\theta = \pi/6$, so the polar coordinates are $(2,\pi/6)$.

(d) $r = \sqrt{3 + (-1)^2} = 2$; $\cos \theta = \sqrt{3}/2$ implies $\theta = \pm\pi/6$, but $\sin \theta =$
-1/2 implies $\theta = -\pi/6$, so the polar coordinates are $(2,-\pi/6)$.

(e) $r = \sqrt{(-\sqrt{3})^2 + (1)^2} = 2$; $\cos \theta = -\sqrt{3}/2$ implies $\theta = \pm 5\pi/6$, but
$\sin \theta = 1/2$ implies $\theta = 5\pi/6$, so the polar coordinates are
$(2,5\pi/6)$.

21. Use the formulas $x = r \cos \theta$ and $y = r \sin \theta$ to convert from polar
coordinates to cartesian coordinates.

(a) $x = 6 \cos(\pi/2) = 6(0) = 0$; $y = 6 \sin(\pi/2) = 6(1) = 6$, so the
cartesian coordinates are (0,6) .

21. (b) $x = -12 \cos(3\pi/4) = -12(-\sqrt{2}/2) = 6\sqrt{2}$; $y = -12 \sin(3\pi/4) = -12(\sqrt{2}/2) =$

$-6\sqrt{2}$,   so the cartesian coordinates are $(6\sqrt{2}, -6\sqrt{2})$.

(c) $x = 4 \cos(-\pi) = 4(-1) = -4$; $y = 4 \sin(-\pi) = 4(0) = 0$,   so the

cartesian coordinates are $(-4,0)$.

(d) $x = 2 \cos(13\pi/2) = 2 \cos(\pi/2) = 2(0) = 0$; $y = 2 \sin(13\pi/2) =$

$2 \sin(\pi/2) = 2(1) = 2$,   so the cartesian coordinates are $(0,2)$.

(e) $x = 8 \cos(-2\pi/3) = 8(-1/2) = -4$; $y = 8 \sin(-2\pi/3) = 8(-\sqrt{3}/2) =$

$-4\sqrt{3}$,   so the cartesian coordinates are $(-4, -4\sqrt{3})$.

(f) $x = -1 \cos(2) = 0.42$; $y = -1 \sin(2) = -0.91$,   so the cartesian

coordinates are $(0.42, -0.91)$.

25.

Referring to the figure, $\cos \theta = x/1$ and

$\cos(-\theta) = x/1$ also; therefore, $\cos \theta = \cos(-\theta)$.

29. By the definition of sine, we have $8/c = \sin(\pi/4) = \sqrt{2}/2$.   Therefore,

$c = 8/(\sqrt{2}/2) = 16/\sqrt{2} = 8\sqrt{2}$.

33.

Let the height be h meters. Then $\tan 17° =$

$h/3000$ implies $h = 3000 \tan 17° \approx 917.19$ meters.

37. Use the law of cosines: $c^2 = a^2 + b^2 - 2ab \cos \theta$ where θ is the measure

of the angle (in radians) opposite c . Let x be the length of the un-

known side. It is determined by $10^2 = x^2 + 5^2 - 2(x)(5)\cos(4\pi/5)$ or

$0 = x^2 + 8.09x - 75$. Thus, the quadratic formula gives $x \approx 13.60$. Now,

the law of cosines yields $5^2 = 10^2 + (13.60)^2 - 2(10)(13.60)\cos r$ or

$\cos r = 0.956$, i.e., $r = 0.30$ radians.

41. From the text, we get $\cos(\theta - \phi) = \cos\phi \cos\theta + \sin\phi \sin\theta$ and $\cos(\theta + \phi) = \cos\theta \cos\phi - \sin\theta \sin\phi$. Subtract to get $\cos(\theta - \phi) - \cos(\theta + \phi) = 2\sin\theta\sin\phi$. Division by two gives the desired product formula.

45. By the addition formula, $\cos(3\pi/2 - \theta) = \cos(3\pi/2)\cos\theta + \sin(3\pi/2)\sin\theta = 0\cdot\cos\theta + (-1)\sin\theta = -\sin\theta$.

49. Using the addition formulas, we get $\cos(\theta + \pi/2)\sin(\phi - 3\pi/2) = [\cos\theta \cos(\pi/2) - \sin\theta \sin(\pi/2)]\cdot[\sin\phi \cos(3\pi/2) - \cos\phi \sin(3\pi/2)] = [0\cdot\cos\theta - 1\cdot\sin\theta]\cdot[0\cdot\sin\phi - (-1)\cos\phi] = -\sin\theta\cos\phi$.

53. Use the half-angle formula. $\cos(\pi/12) = \cos[(\pi/6)/2] = \sqrt{[1 + \cos(\pi/6)]/2} = \sqrt{(1 + \sqrt{3}/2)/2} = \sqrt{2 + \sqrt{3}}/2$. Now $\sec(\pi/12) = 1/\cos(\pi/12) = 2/\sqrt{2 + \sqrt{3}}$.

57. Take the reciprocal of both sides and use the half-angle formula $\cos^2(\theta/2) = (1 + \cos\theta)/2$. Then $\sec^2(\theta/2) = (2\sec\theta)/(\sec\theta + 1)$ implies $\cos^2(\theta/2) = (\sec\theta + 1)/(2\sec\theta) = 1/2 + 1/(2\sec\theta) = 1/2 + (\cos\theta)/2 = (1 + \cos\theta)/2 = \cos^2(\theta/2)$.

61.

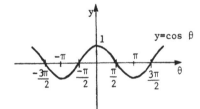

We obtain $y = 2\cos 3\theta$ by compressing the graph of $y = \cos\theta$ horizontally by a factor of 3 and stretching it vertically by a factor of 2.

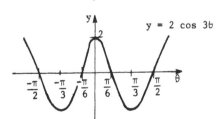

65.

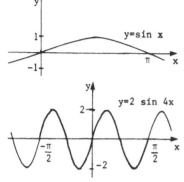

We begin by simplifying $y = 4 \sin 2x \times$
$\cos 2x = 2(2 \sin 2x \cos 2x) =$
$2(\sin 2(2x)) = 2 \sin 4x$. Hence, we
obtain $y = 4 \sin 2x \cos 2x$ by compressing the graph of $y = \sin x$ horizontally by a factor of 4 and stretching it vertically by a factor of 2 .

69. From Fig. 5.1.4(f), notice that the concavity changes at $-3\pi/2$, $-\pi/2$,
$\pi/2$, and $3\pi/2$. In general, the inflection points for $\cot \theta$ are
$(2n + 1)\pi/2$ where n is an integer.

73. (a) Rearrangement of Snell's law yields $v_1(\sin \theta_2/\sin \theta_1) = v_2 =$
$(3 \times 10^{10})(\sin(30°)/\sin(60°)) = (3 \times 10^{10})[(1/2)/(\sqrt{3}/2)] =$
$(3 \times 10^{10})/\sqrt{3} = \sqrt{3} \times 10^{10}$ cm/sec .

(b) If $v_1 = v_2$, then $\sin \theta_1 = \sin \theta_2$ implies $\theta_1 = \theta_2$. Therefore
the angle of incidence is the same as the angle of refraction, so
the light travels in a straight line.

(c) We have $v_2 = v_1/2$, so $v_1/v_2 = 2 = \sin(45°)/\sin \theta_2 = (\sqrt{2}/2)/$
$\sin \theta_2$. This implies $\sin \theta_2 = \sqrt{2}/4$, so $\theta_2 = 0.36$ radians or
$20.7°$.

77.

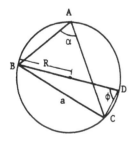

We will show that $\sin \alpha /a = 1/2R$ where
R is the radius of the circumscribed circle.
Let $\overline{BD}$ be a diameter of the circle, then
geometry tells us that $\angle BCD$ is a right angle.
Therefore, $\sin \phi = a/2R$. Angle α and
angle ϕ subtend the same arc, so $\alpha = \phi$ and
$\sin \alpha = a/2R$. Consequently, $\sin \alpha/a = 1/2R$.

81. (a) When $\sin \theta = \lambda/a$, we have $I = I_0 \{ \sin [\pi a (\lambda/a)/\lambda] / [\pi a (\lambda/a)/\lambda] \}^2 = I_0 (\sin \pi/\pi)^2 = I_0 (0/\pi)^2 = 0$.

   (b) $\sin \phi = 0$ for $\phi = n\pi$, where n is an integer; therefore, we solve $\pi a \sin \theta/\lambda = n\pi$. The solution is $\sin \theta = n\lambda/a$, so θ is all positive values whose sine is $n\lambda/a$.

   (c) When $\pi a \sin \theta/\lambda$ is close to zero, Exercise 74 tells us that the numerator of the squared term is approximately equal to its denominator, so the squared term is approximately 1 . Therefore, I is approximately I_0 .

   (d) $\sin \theta = (5 \times 10^{-5})/(10^{-2}) = 5 \times 10^{-3}$ implies $\theta = 5.000021 \times 10^{-3} \approx 5 \times 10^{-3} = \lambda/a$.

SECTION QUIZ

1. $(-4,4)$ is a point described in cartesian coordinates. What are its corresponding polar coordinates?

2. Suppose $(r,\theta) = (\pi/4,1)$ are the polar coordinates of a point.

   (a) Use a negative radius to describe the same point.

   (b) Convert to cartesian coordinates, rounding to the nearest 0.1 .

3. Let $(3,-4)$ be the cartesian coordinates of a point and let (r,θ) be the polar coordinates. Without using a calculator or tables, find:

   (a) $\sin \theta$

   (b) $\sec \theta$

4. Sketch $y = \cos x$. Trace the graph and relabel the axes to depict $y = (1/3)\cos 2x$.

5. To celebrate the end of another midterm, Merry May decided to throw a
    party. Since polar coordinates were still on her mind, she located the
    beginning of a song on a record at the polar coordinates A(5,-π/4);
    the song ends at the polar coordinates C(3,3π/4) .   Another song begins
    at point B which is -3 units from A at an angle of -π/6 .

    (a) What are the cartesian coordinates of A , B , and C ?

    (b) Suppose Merry May only had a straightedge to help her locate songs.
         Plot all three points and help her determine the distance from C
         to B .

ANSWERS TO SECTION QUIZ

1. $(4\sqrt{2}, 3\pi/4)$

2.   (a) $(-\pi/4, 1 + \pi)$

     (b)   (0.4, 0.7)

3.   (a)   -4/5

     (b)   5/3

4.

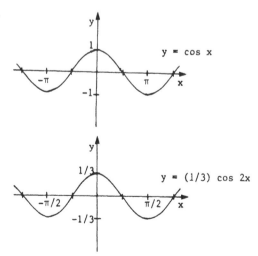

5.   (a) $A(5\sqrt{2}/2, -5\sqrt{2}/2)$

$B((5\sqrt{2} - 3\sqrt{3})/2, (-5\sqrt{2} + 3)/2)$

$C(-3\sqrt{2}/2, 3\sqrt{2}/2)$

(b) $[73 - 12\sqrt{6} - 12\sqrt{2}]^{1/2}$

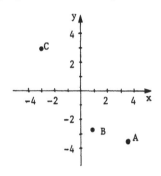

5.2 Differentiation of the Trigonometric Functions

PREREQUISITES

1. Recall how to define the trigonometric functions in terms of sines and
    cosines (Section 5.1).

2. Recall the definition of the derivative (Section 1.3).

3. Recall the basic rules of differentiation (Sections 1.4, 1.5, and 2.2).

4. Recall how to compute an antiderivative and how to use the fundamental
    theorem of calculus (Sections 2.5 and 4.4).

PREREQUISITE QUIZ

1. Express the following in terms of sin x and cos x :

    (a) tan x

    (b) cot x

    (c) sec x

    (d) csc x

2. What is the relationship between $f(x_0)$, Δx , and $f'(x_0)$?

3. Differentiate the following functions:

    (a) $(x^2 + 2x + 2)(x^2 + 3x)$

    (b) $(x^2 + 2x + 2)/(x^2 + 3x)$

    (c) $(x^3 + 3x - 5)^5$

4. State the fundamental theorem of calculus.

5. Suppose $(d/dx)\Delta(x) = \delta(x)$; what is $\int \delta(x)dx$?

GOALS

1. Be able to differentiate trigonometric functions.

2. Be able to integrate certain trigonometric functions.

STUDY HINTS

1.   Derivative of sine, cosine. For the purposes of this course, you should
     memorize that the derivative of $\sin x$ and $\cos x$ are $\cos x$ and
     $-\sin x$, respectively. It is not essential to know the derivation.
     Don't forget the minus sign in $(d/dx)\cos x = -\sin x$.

2.   Other derivatives of trigonometric functions. You may find it easier to
     write the other trigonometric functions in terms of $\sin$ and $\cos$ and
     then use the rules of differentiation to derive the derivatives. On the
     other hand, you may prefer to memorize the differentiation formulas. One
     can simply memorize the derivatives of $\tan$ and $\sec$ and then remember
     that the derivatives of the cofunctions are the negative cofunctions. For
     example, we know that $(d/dx)\tan x = \sec^2 x$. Then, since $\cot x$ and
     $\csc x$ are the cofunctions of $\tan x$ and $\sec x$, respectively, we get
     $(d/dx)\cot x = -\csc^2 x$.

3.   Trigonometric antiderivatives. There is really nothing new to learn, but
     it is important to get used to the notation and to reading the differentia-
     tion formulas backwards. The antidifferentiation formulas follow directly
     from the differentiation formulas derived in this section.

SOLUTIONS TO EVERY OTHER ODD EXERCISE

1.   By the sum rule, $(d/d\theta)(\cos\theta + \sin\theta) = -\sin\theta + \cos\theta$.

5.   By the product rule, $(d/d\theta)[(\cos\theta)(\sin\theta + \theta)] = (-\sin\theta)(\sin\theta + \theta) +$
     $(\cos\theta)(\cos\theta + 1) = \cos^2\theta - \sin^2\theta - \theta\sin\theta + \cos\theta = \cos 2\theta - \theta\sin\theta +$
     $\cos\theta$.

9.   By the quotient rule, $(d/d\theta)[\cos\theta/(\cos\theta - 1)] = [(-\sin\theta)(\cos\theta - 1) -$
     $\cos\theta(-\sin\theta)]/(\cos\theta - 1)^2 = \sin\theta/(\cos\theta - 1)^2$.

13. By the chain rule, $(d/dx)(\cos x)^3 = 3(\cos x)^2(-\sin x) = -3 \cos^2 x \sin x$.

17. By the chain rule, $(d/dx)\sin(x + \sqrt{x}) = [\cos(x + \sqrt{x})](1 + 1/2\sqrt{x})$.

21. By the sum rule, $(d/dx)(\tan x + 2 \cos x) = \sec^2 x - 2 \sin x$.

25. $df(x)/dx = 1/2\sqrt{x} - 3 \sin 3x$ by applying the chain rule to $\cos 3x$.

29. $df(t)/dt = (12t^2)\sin\sqrt{t} + (4t^3 + 1)(\cos\sqrt{t})(1/2\sqrt{t}) = 12t^2\sin\sqrt{t} +$
    $[(4t^3 + 1)/2\sqrt{t}]\cos\sqrt{t}$. The chain rule and the product rule were used.

33. Apply the chain rule and quotient rule to get $df(\theta)/d\theta =$
    $(3/2)[\csc(\theta/\sqrt{\theta^2 + 1}) + 1]^{1/2}[-\cot(\theta/\sqrt{\theta^2 + 1})\csc(\theta/\sqrt{\theta^2 + 1})] \cdot \{[\sqrt{\theta^2 + 1} -$
    $\theta(1/2)(\theta^2 + 1)^{-1/2}(2\theta)]/(\theta^2 + 1)\} = -3[\csc(\theta/\sqrt{\theta^2 + 1}) + 1]^{1/2}[\cot(\theta/$
    $\sqrt{\theta^2 + 1})][\csc(\theta/\sqrt{\theta^2 + 1})]/2(\theta^2 + 1)^{3/2}$.

37. $\int(x^3 + \sin x)dx = x^4/4 - \cos x + C$.

41. Guess that the antiderivative is $a \cos(u/2)$. Differentiation gives
    $-(1/2)a \sin(u/2)$, so $a = -2$; therefore $\int\sin(u/2)du = -2 \cos(u/2) + C$.

45. The useful trigonometric identity is $\sin 2\theta = 2\sin \theta \cos \theta$, so
    $\int\cos \theta \sin \theta d\theta = (1/2)\int\sin 2\theta d\theta$. Guess that the antiderivative has the
    form $a \cos 2\theta$. Differentiation gives $-2a \cos 2\theta$, so $a = -1/2$.
    Therefore, the integral is $(1/2)(-1/2) \cos 2\theta + C = -\cos 2\theta/4 + C$.

49. The antiderivative should have the form $a \cos(\theta/4)$. Differentiation
    gives $-(a/4)\sin(\theta/4)$, so $a = -4$; therefore, $\int_0^{\pi/2}\sin(\theta/4)d\theta =$
    $-4 \cos(\theta/4)\big|_0^{\pi/2} = 4 - 4 \cos (\pi/8)$.

53. Differentiating $a \sin(3\pi t)$ gives $3\pi a \cos(3\pi t)$. This implies that
    $\int_0^1\cos(3\pi t)dt = \sin(3\pi t)/3\pi\big|_0^1 = (\sin 3\pi - \sin 0)/3\pi = 0$.

57. (a) Note that $\sin 2\phi/(1 + \cos 2\phi) = 2 \sin \phi \cos \phi/(1 + 2 \cos^2\phi - 1) =$
    $\sin \phi/\cos \phi = (\sin \phi/\phi) \cdot (\phi/\cos \phi)$. Since $\cos \phi < \sin \phi/\phi < 1$
    for $0 < \phi < \pi/2$, we have $\cos \phi \cdot (\phi/\cos \phi) < (\sin \phi/\phi) \cdot (\phi/\cos \phi)$.
    That is, $\phi < (\sin \phi/\phi) \cdot (\phi/\cos \phi) = \sin 2\phi/(1 + \cos 2\phi)$ for $0 < \phi <$
    $\pi/2$.

57. (b) Note that $\sec \phi = 1/\cos \phi$. Since $\cos \phi < \sin \phi/\phi$ and $\sin \phi <$
1 for $0 < \phi < \pi/2$, $\cos \phi < 1/\phi$, i.e., $1/\cos \phi > \phi$. Hence
$\sec \phi = 1/\cos \phi > \phi$.

61. Let $f(x) = \sin x$, then $f'(x) = \cos x$ and $f''(x) = -\sin x$, so
$f''(x) + f(x) = 0$. Let $f(x) = \cos x$, then $f'(x) = -\sin x$ and
$f''(x) = -\cos x$, so $f''(x) + f(x) = 0$.

65. Differentiate by using the chain rule. $(d/d\theta)[-f(\cos \theta)] = -f'(\cos \theta) \times$
$[(d/d\theta)\cos \theta]$. Since $f'(x) = 1/x$, we have $-(1/\cos \theta)(-\sin \theta) =$
$\tan \theta$. Thus, $-f(\cos \theta) + C$ is the antiderivative for $\tan \theta$.

69. (a) Using the product rule, $(d/dx)[\phi(3x)\cos x] = \phi(3x)\cdot(-\sin x) +$
$(\cos x)\phi'(3x)3 = -(\sin x)\cdot\phi(3x) + 3\cos x/\cos(3x)$.

(b) Since $d\phi/dx = 1/\cos x$, an antiderivative for $1/\cos x$ is $\phi(x)$.
Thus, $\int_0^1 (dx/\cos x) = \phi(x)\big|_0^1 = \phi(1) - \phi(0)$.

(c) Using the product rule, $(d/dx)(\phi(2x)\sin 2x) = 2\phi'(2x)\cdot\sin 2x + \phi(2x) \times$
$(2\cos 2x) = 2\sin 2x/\cos 2x + 2\phi(2x)\cos 2x$. Differentiating again,
$(d^2/dx^2)(\phi(2x)\sin 2x) = [4\cos 2x(\cos 2x) - 2\sin 2x(-2\sin 2x)]/\cos^2 2x +$
$2\cdot 2\phi'(2x)\cos 2x + 2\phi(2x)\cdot(-2\sin 2x) = 4(\cos^2 2x + \sin^2 2x)/\cos^2 2x +$
$4\cos 2x/\cos 2x - 4\phi(2x)\cdot\sin 2x = 4\sec^2 2x + 4 - 4\phi(2x)\cdot\sin 2x$.

SECTION QUIZ

1. What is wrong with this statement? If $f(x) = \sin^3 3x$, then $f'(x) =$
$3 \sin^2 3x \cos 3x$.

2. Differentiate the following with respect to x :

(a) $\cos x \sin x$

(b) $\sec(\pi/4)$

(c) $\tan(x^3)$

(d) $\sec^2 x$

3.   Evaluate the following integrals:

(a) $\int_0^\pi \sin 3x \, dx$

(b) $\int_0^{\pi/2} \cos \pi \, dx$

4.   As usual, the Do-Wrong Construction Company did wrong again. Just
     yesterday, it finished building a storage room for nuclear reactor
     parts; however, Do-Wrong forgot to make a door. Due to expense, it
     is desirable to remove as little wall material as possible. The
     largest nuclear reactor part has the shape of the region under $y =$
     $\tan^2 x$ for $\pi/6 \leqslant x \leqslant \pi/3$. How much of the wall needs to be removed?
     (Hint: $\tan^2 \theta + 1 = \sec^2 \theta$.)

ANSWERS TO PREREQUISITE QUIZ

1.   (a) $\sin x / \cos x$

     (b) $\cos x / \sin x$

     (c) $1/\cos x$

     (d) $1/\sin x$

2. $f'(x_0) = \lim_{\Delta x \to 0} \{ [f(x_0 + \Delta x) - f(x_0)] / \Delta x \}$

3.   (a) $(2x + 2)(x^2 + 3x) + (x^2 + 2x + 2)(2x + 3)$

     (b) $[(2x + 2)(x^2 + 3x) - (x^2 + 2x + 2)(2x + 3)] / (x^2 + 3x)^2$

     (c) $15(x^3 + 3x - 5)^4 (x^2 + 1)$

4.   If $F'(x) = f(x)$, then $\int_a^b f(x) \, dx = F(x) \big|_a^b$.

5. $\Delta(x) + C$

ANSWERS TO SECTION QUIZ

1. $f'(x) = 9 \sin^2 3x \cos 3x$;   9   comes from the exponent   3   and   the
     derivative of $3x$.

2.   (a) $\cos^2 x - \sin^2 x$

    (b)   0

    (c) $3x^2 \sec^2(x^3)$

    (d) $2 \sec^2 x \tan x$

3.   (a)   2/3

    (b) $-\pi/2$

4. $2\sqrt{3}/3 - \pi/6$

5.3   Inverse Functions

PREREQUISITES

1. Recall how to use the vertical line test (Section R.6).

2. Recall how to use the intermediate value theorem (Section 3.1).

3. Recall how to evaluate composite functions and how to differentiate
    them (Section 2.2).

4. Recall the definition of the derivative (Section 1.3).

PREREQUISITE QUIZ

1. Suppose $f(x) = x^8$, what is $\lim_{\Delta x \to 0} \{[f(x + \Delta x) - f(x)]/\Delta x\}$?

2. If $f(0) = 1$ and $f(2) = -2$, and f is continuous, can you locate
    a root of f ? Explain.

3. Let $f(x) = x^2$ and $g(x) = \cos x$.

    (a) Find $f(g(x))$ and its derivative.

    (b) Find $g \circ f(x)$ and its derivative.

4. Explain the vertical line test.

GOALS

1. Be able to state and explain the definition of an inverse.

2. Be able to differentiate simple inverse functions.

STUDY HINTS

1. Inverses. For most cases, we express a function by $y = f(x)$. If it
    is possible to uniquely find x in terms of y , then we say that x
    is the inverse of y . However, in many instances, an inverse may
    exist even if we cannot explicitly find x in terms of y . The graphs

1. Inverses (continued).

    of inverses may be obtained by "flipping" the axes. See Fig. 5.3.1.

2. Horizontal line test. If each horizontal line meets the graph of f
    in at most one point, then f is invertible. Compare how this state-
    ment corresponds to the vertical line test (See Section R.6). How does
    "flipping" axes make this statement plausible?

3. Inverse function test. All the test says is that any part of a function
    that is strictly increasing or strictly decreasing is invertible. Think
    about what problem exists if the function is not strictly increasing or
    strictly decreasing. The domain is $[f(a),f(b)]$ or $[f(b),f(a)]$; just
    be sure the smaller number comes first.

4. Inverse function rule. Probably the easiest formula to remember is
    $dx/dy = 1/(dy/dx)$ because the differentials act like regular fractions.
    The formula is very easy to use if you are differentiating at a specific
    point (See Example 8). If you want to differentiate an inverse function,
    you need to know both the function and its inverse. The best way to
    learn to use the formula is to study Example 7 and practice. (You will
    get more practice in the next section.)

SOLUTIONS TO EVERY OTHER ODD EXERCISE

1. $f'(x) = 2$ does not change sign, so $f(x)$ has an inverse. We have
    $f(x) = y = 2x + 5$, $f(-4) = -3$, and $f(4) = 13$. Solving for the
    independent variable in terms of the dependent variable, we get x =
    $(y - 5)/2$. Changing the names of the variables yields $f^{-1}(x) =$
    $(x - 5)/2$ on $[-3,13]$.

5. $h'(t) = [(t + 3) - (t - 10)]/(t + 3)^2 = 13/(t + 3)^2$ does not change

sign, so $h(t)$ has an inverse.   We have $h(t) = (t - 10)/(t + 3)$,

$h(-1) = -11/2$,   and $h(1) = -9/4$.   Solving for the independent

variable in terms of the dependent variable, we get $ht + 3h = t - 10$

or $t(h - 1) = -3h - 10$.   Thus, the inverse is $t = (3h + 10)/(1 - h)$

or $h^{-1}(t) = (3t + 10)/(1 - t)$ on $[-11/2,-9/4]$.

9.   (a)

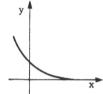

(b)

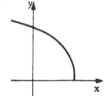

(c)

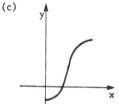

The inverses were graphed by tracings
through the back of the page and then
rotating the x-axis into a vertical
position.

13.

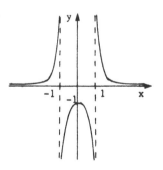

$f'(x) = -(x^2 - 1)^{-2} \cdot 2x$,   so the critical
point is $x = 0$.   Thus,   f   is increasing
on $(-\infty,-1)$ and $(-1,0)$.   It is decreasing
on $(0,1)$ and $(1,\infty)$.   From the graph at
the left, we see that the largest interval
on which   f   is invertible is $(-\infty,-1)$ or
$(1,\infty)$.   The graphs of the inverses are

shown below and were obtained by flipping through the line $x = y$.

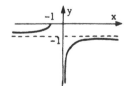

17. We apply the inverse function test. $f'(x) = x^2 - 1$, which is zero if
$x = \pm 1$. f is increasing on $(-\infty,-1)$ and $(1,\infty)$, and decreasing on
$(-1,1)$. Hence f is not strictly increasing or decreasing on any
open interval containing 1 . That is, f is not invertible on such
an interval.

21. $f'(x) = -3x^2 - 2$, which is negative for all x . Thus, f is decreasing on $[-1,2]$. Also, f is continuous, so by the inverse function test, f is invertible on $[-1,2]$. The domain of the inverse is
$[f(2),f(-1)]$, i.e., $[-11,4]$. The domain is $[f(b),f(a)]$ since f
is decreasing.

25. Since f is a polynomial, it is continuous. The derivative is $f'(x) =$
$3x^2 + 2$, which is always positive. By the inverse function test, f
has an inverse $g(y)$. Notice that $g(4) = 1$ since $f(1) = 4$. From
the inverse function rule, with $y_0 = 4$ and $x_0 = 1$, $g'(4) = 1/f'(1) =$
$1/[3(1)^2 + 2] = 1/5$.

29. $f'(x) = 3$ for all x , so by the inverse function rule, $g'(y) = 1/f'(x) =$
$1/3$ for all x . Thus, we have $g'(2) = g'(3/4) = 1/3$.

33. The errors do not seem to build up on most calculators. In most cases,
you should get back $|x|$, which is not the original number if you
started with a negative number. This demonstration does not work if
x is too small or too large. Notice that $\sqrt{x}$ and x^2 are inverses
only if $x \geq 0$.

37. (a) Suppose $3x + 7 > 0$ or $x > -7/3$, then $2x + 5 \geq 0$ or $x \geq -5/2$;
therefore, we must have $x > -7/3$. Now, if $3x + 7 < 0$ or $x <$
$-7/3$, then $2x + 5 \leq 0$ or $x \leq -5/2$; therefore, we must have
$x \leq -5/2$. Thus, the domain is $(-\infty,-5/2]$ and $(-7/3,+\infty)$.

37. (b) An equivalent expression is $y^2 = (2x + 5)/(3x + 7)$, which implies $3xy^2 + 7y^2 = 2x + 5$ or $x(3y^2 - 2) = 5 - 7y^2$. Therefore, the inverse function is $g(y) = (5 - 7y^2)/(3y^2 - 2)$.

(c) $f'(x) = 1/g'(y) = \{[(-14y)(3y^2 - 2) - (5 - 7y^2)(6y)]/(3y^2 - 2)^2\}^{-1} =$
$-(3y^2 - 2)^2/2y = -[3((2x + 5)/(3x + 7)) - 2]^2/2\sqrt{(2x + 5)(3x + 7)} =$
$-[1/(3x + 7)^2]/2\sqrt{(2x + 5)(3x + 7)} = -1/[2(3x + 7)^{3/2}\sqrt{2x + 5}]$.

(d) $f'(x) = (1/2)[(2x + 5)/(3x + 7)]^{-1/2}[(2)(3x + 7) - (2x + 5)(3)]/$
$(3x + 7)^2 = -1/[2(3x + 7)^{3/2}\sqrt{2x + 5}]$.

41. (a) By the definition of the derivative, $f'(x_0) = \lim_{\Delta x \to 0}(\Delta y/\Delta x)$. Thus, as Δx gets smaller, $\Delta y/\Delta x$ approaches $f'(x_0)$, which is positive according to the hypothesis. Therefore, we have $(3/2)f'(x_0) > \Delta y/\Delta x > (1/2)f'(x_0)$ for Δx sufficiently small.

(b) Let us assume Δx does not approach 0 ; then $\Delta y/\Delta x$ must approach 0 as Δy approaches 0 . This cannot occur because $\Delta y/\Delta x$ is between two positive numbers.

(c) $\Delta x = x - x_0 = g(y_0 + \Delta y) - g(y_0)$. Then $g(y_0 + \Delta y) = g(y_0) + \Delta x$ approaches $g(y_0)$ as Δy approaches zero since $\Delta x \to 0$. Thus g is continuous at y_0 .

SECTION QUIZ

1. Suppose $f(x) = (x^3 + 2x + 8)/x$. On what intervals is f invertible?

2. For the function in Question 1, one of the intervals contains $x = 1$. What is the derivative of f^{-1} at that point?

3. Find a formula for f^{-1} if $f(x) = (x - 1)/(x + 5)$.

4.   Backwards Billy was born with a neurological disease which caused him
     to think backwards. When asked to sketch the graph of $y = x^4 - 3x^3 +$
     $x^2 + 8$, he wants to start by sketching the inverse and then flipping
     the graph. However, you realize that y doesn't have an inverse.

     (a) Explain to Backwards Billy why the inverse doesn't exist.

     (b) Find the largest interval around $x = 1$ for which an inverse
          exists.

     (c) On the restricted interval, what is the derivative of the inverse
          at $x = 1$?

ANSWERS TO PREREQUISITE QUIZ

1. $8x^7$

2.   By the intermediate value theorem, there is at least one root in $(0,2)$.

3. $f(g(x)) = \cos^2 x$ and $f'(g(x)) = -2 \sin x \cos x$; $g \circ f(x) = \cos (x^2)$
     and $g'(f(x)) = -2x \sin (x^2)$.

4.   If a vertical line drawn at each value of x intersects a graph at only
     one point, then it is the graph of a function.

ANSWERS TO SECTION QUIZ

1. $(-\infty,0)$, $(0,\sqrt[3]{4})$, and $(\sqrt[3]{4},\infty)$

2. $-1/6$

3. $f^{-1}(x) = (5x + 1)/(1 - x)$

4.   (a) The inverse function test is violated; $f'(x) > 0$ on $(0,1/4)$
          and $(2,\infty)$ and $f'(x) < 0$ on $(-\infty,0)$ and $(1/4,2)$.

     (b) $[1/4,2]$

     (c) $-1/3$

5.4   The Inverse Trigonometric Functions

PREREQUISITES

1.   Recall the concept of an inverse function (Section 5.3).

2.   Recall how to differentiate an inverse function (Section 5.3).

3.   Recall how to differentiate trigonometric functions (Section 5.2).

PREREQUISITE QUIZ

1.   Evaluate the following:

   (a) $(d/dx) \sin x$

   (b) $(d/dy) \tan y$

   (c) $(d/dx)f^{-1}(x)|_2$ if $f(x) = x^3 - x$

2.   On what intervals does $f(x) = x^4 + x^2$ have an inverse?

3.   If $g(u) = u^3 + u$,   what is $g^{-1}(2)$?

GOALS

1.   Be able to differentiate the inverse trigonometric functions.

2.   Be able to simplify trigonometric functions of inverses, i.e., what is
   $\tan (\sin^{-1}x)$?

3.   Be able to integrate certain algebraic expressions that lead to inverse
   trigonometric functions.

STUDY HINTS

1.   <u>Notation</u>. In this text and most others, $\sin^{-1}x$ is the inverse function
   of $\sin x$, not $1/\sin x$. Do not confuse this notation with $\sin^2 x$ or
   $\sin^3 x$, which mean $(\sin x)^2$ and $(\sin x)^3$, respectively. The notation
   arcsin, arccos, etc., is also commonly used to designate inverse trigonometric functions.

2.   **Inverse sine.** Remember that angles are expressed in radians. We have
     chosen x to be in $[-\pi/2, \pi/2]$. Since $x = \sin^{-1}y$ means $\sin x = y$,
     note that y must be in $[-1,1]$.

3.   **Trigonometric functions of inverses.** Be sure you understand Example
     2(b). By drawing an arbitrary right triangle and labelling it appro-
     priately by using the definitions of the trigonometric functions and
     Pythagoras' theorem, one can easily find the other trigonometric functions.

4.   **Range for which inverses exist.** The angles for which the trigonometric
     functions have inverses include those in the first quadrant of the xy-
     plane and the adjacent (second or fourth) quadrant in which the function
     is negative. If there are two choices, then choose the quadrant which
     maintains continuity.

5.   **Derivatives.** You may desire to learn to derive the derivatives of the
     inverses by using the inverse function rule, but it can be time consuming.
     If memorizing is your forte, you only need to know the derivatives of
     $\sin^{-1}y$, $\tan^{-1}y$, and $\sec^{-1}y$. The derivatives of the inverses of the
     cofunctions are the same except for a sign change.

6.   **Derivative of $\sec^{-1}x$, $\csc^{-1}x$.** Notice that the donominator is a square
     root which is positive. Thus, $\sqrt{y^2(y^2 - 1)}$ simplifies to $|y|\sqrt{y^2 - 1}$.
     The absolute value sign is a necessity. See Example 8.

7.   **Antiderivatives.** In formula (4) on p. 287, we are given $\int [dx/(1 + x^2)] =$
     $\tan^{-1}x + C$. Another valid formula is $\int [dx/(1 + x^2)] = -\cot^{-1}x + C'$.
     These formulas are equivalent since the arbitrary constants, C and
     C', may differ.

SOLUTIONS TO EVERY OTHER ODD EXERCISE

1. Since $\sin(0.615) \approx 1/\sqrt{3}$, $\sin^{-1}(1/\sqrt{3}) \approx 0.615$. Remember that the domain of $\sin^{-1}x$ is $[-\pi/2, \pi/2]$.

5. Since $\cos(0) = 1$, $\cos^{-1}(1) = 0$. Remember that the domain of $\cos^{-1}x$ is $[0, \pi]$.

9. $\sec^{-1}(2/\sqrt{3})$ is the same as $\cos^{-1}(\sqrt{3}/2)$, so since $\cos(\pi/6) = \sqrt{3}/2$, $\sec^{-1}(2/\sqrt{3}) = \pi/6$. The domain of $\sec^{-1}x$ is $[0, \pi/2)$ and $(\pi/2, \pi]$.

13. Use the product rule along with the fact that $(d/dx)\sin^{-1}x = 1/\sqrt{1 - x^2}$. Thus, the derivative is $2x \sin^{-1}x + x^2/\sqrt{1 - x^2}$.

17. Combine the chain rule and the quotient rule along with the fact that $(d/dx) \tan^{-1}x = 1/(1 + x^2)$. Thus, the derivative is $[1 + ((2x^5 + x)/(1 - x^2))^2]^{-1} \cdot (d/dx)[(2x^5 + x)/(1 - x^2)] = [(1 - x^2)^2/((1 - x^2)^2 + (2x^5 + x)^2)] \cdot [(1 - x^2)(10x^4 + 1) - (2x^5 + x)(-2x)]/(1 - x^2)^2 = [(1 - x^2)(10x^4 + 1) + 2x^2(2x^4 + 1)]/[(1 - x^2)^2 + x^2(2x^4 + 1)^2]$.

21. Use the quotient rule along with the fact that $(d/dx) \sin^{-1}x = 1/\sqrt{1 - x^2}$. Thus, the derivative is $[(3/\sqrt{1 - 9x^2})(x^2 + 2) - (\sin^{-1}3x) \times (2x)]/(x^2 + 2)^2 = [3(x^2 + 2) - (2x)(\sin^{-1}3x)\sqrt{1 - 9x^2}]/(x^2 + 2)^2\sqrt{1 - 9x^2}$.

25. Apply the quotient rule to get $(d/dx)[(\cos^{-1}x)/(1 - \sin^{-1}x)] = [(-1/\sqrt{1 - x^2})(1 - \sin^{-1}x) - (-1)(1/\sqrt{1 - x^2})(\cos^{-1}x)]/(1 - \sin^{-1}x)^2 = (\sin^{-1}x + \cos^{-1}x - 1)/(1 - \sin^{-1}x)^2\sqrt{1 - x^2}$.

29. Since $(d/dx) \tan^{-1}x = (1 + x^2)^{-1}$, we get $\int [3/(1 + x^2) + x]\,dx = 3 \tan^{-1}x + x^2/2 + C$.

33. $\int [3/(1 + 4x^2)]\,dx = \int [3/(1 + (2x)^2)]\,dx$, so we expect an antiderivative of the form $a[\tan^{-1}(2x)]$. Differentiation implies that $a = 3/2$, so the answer is $3 \tan^{-1}(2x)/2 + C$.

37.

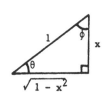

As shown in the diagram, $\sin^{-1}x = \theta$; therefore,

$\tan(\sin^{-1}x) = \tan\theta = x/\sqrt{1 - x^2}$.

41. Since $(d/d\theta)\cos^{-1}\theta = -1/\sqrt{1 - \theta^2} = -(1 - \theta^2)^{-1/2}$, we have

$(d^2/d\theta^2)\cos^{-1}\theta = -(-1/2)(1 - \theta^2)^{-3/2}(-2\theta) = -\theta/(1 - \theta^2)^{3/2}$.

45. The rate of change is $(d/ds)\cos^{-1}(8s^2 + 2) = [-1/\sqrt{1 - (8s^2 + 2)^2}](16s)$.

At $s = 0$, the derivative is 0 .

49. Use the inverse function rule, $(d/dy)[f^{-1}(y)] = 1/[(d/dx)f(x)]$. Thus,

$(d/dy)\cot^{-1}y = 1/(d/dx)\cot x = -1/\csc^2 x = -1/(1 + \cot^2 x) = -1/(1 + y^2)$

because $y = \cot x$.

53. (a) The best method is to simplify and then use implicit differentiation.

The equation simplifies to $\sin(x + y) = xy$, which implies

$[\cos(x + y)][1 + (dy/dx)] = y + x(dy/dx)$. Therefore, $[\cos(x + y) - $

$y]/[x - \cos(x + y)] = dy/dx$. (Note: $\cos(x + y) = $

$\sqrt{1 - \sin^2(x + y)} = \sqrt{1 - x^2y^2}$, so $dy/dx = (\sqrt{1 - x^2y^2} - y)/$

$(x - \sqrt{1 - x^2y^2})$.)

(b) The chain rule gives us $dy/dt = (dy/dx)(dx/dt)$, where $dx/dt = $

$[(1)(1 - t^2) - (t)(-2t)]/(1 - t^2)^2 = (1 + t^2)/(1 - t^2)^2$. Therefore,

$dy/dt = [(\cos(x + y) - y)(1 + t^2)]/[(x - \cos(x + y))(1 - t^2)^2]$.

(c) According to the inverse function rule, $dx/dy = 1/(dy/dx)$. From

the chain rule, we get $dx/dt = (dx/dy)(dy/dt)$ where $dy/dt = $

$1/\sqrt{1 - t^2}$. Therefore, $dx/dt = [x - \cos(x + y)]/[\cos(x + y) - $

$y]\sqrt{1 - t^2}$.

(d) $dx/dt = 3t^2 + 2$. By the method of part (b), we have $dy/dt = $

$[\cos(x + y) - y](3t + 2)/[x - \cos(x + y)]$.

57. Begin by simplifying the first term, so $\int [(x^2 + 1)/x^2 + 1/\sqrt{1 - x^2} +$

   $\cos x]\, dx = \int [1 + x^{-2} + 1/\sqrt{1 - x^2} + \cos x]\, dx = x - 1/x + \sin^{-1}x +$

   $\sin x + C$. Evaluating at $\sqrt{2}/2$ and $1/2$, we get $\sqrt{2}/2 - 2/\sqrt{2} +$

   $\sin^{-1}(\sqrt{2}/2) + \sin(\sqrt{2}/2) - 1/2 + 2 - \sin^{-1}(1/2) - \sin(1/2) = [\sqrt{2} -$

   $2\sqrt{2} - 1 + 4]/2 + \pi/4 - \pi/6 + \sin(\sqrt{2}/2) - \sin(1/2) = (3 - \sqrt{2})/2 +$

   $\pi/12 + \sin(\sqrt{2}/2) - \sin(1/2)$.

SECTION QUIZ

1. Explain why $\sin(\sin^{-1}x) = x$ for all x in the domain of $\sin^{-1}x$, but
    $\sin^{-1}(\sin x) = x$ only for some x in the domain of $\sin x$.

2. Simplify:

    (a) $\tan(\cos^{-1}x)$

    (b) $\sin 2\theta$ if $\theta = \tan^{-1}(x/2)$

    (c) $\tan^{-1}(-\cos(-\pi))$

3. Why doesn't $\cos(\sin^{-1}(2 + x^2))$ exist?

4. How would you define $\tan^{-1}(\pi/2)$?

5. Your roommate, who has been frantically studying about inverses, was
    doing the dinner dishes when he imagined seeing little green pea-like
    beings entering through the kitchen sink. He told you that they intro-
    duced themselves, "We're from the Innerverse, the tiny land of inverse
    trigonometric functions." They had just travelled along the path de-
    scribed by $\sin^{-1}(2x - 2)$.

    (a) The pea-like beings wanted to know the steepness of the path at x_0
         for their return trip. Help your roommate get rid of his imaginary
         friends by providing the correct answer.

5.   (b)   Find the width of the path, i.e., determine the length of the

domain of $\sin^{-1}(2x - 2)$.

(c)   Your roommate also mentioned that they might return to find out

the sine, cosine, and tangent of $\sin^{-1}(2x - 2)$.   Compute these

quantities for him to help save his sanity.

ANSWERS TO PREREQUISITE QUIZ

1.   (a) $\cos x$

(b) $\sec^2 y$

(c) $1/11$

2. $(-\infty, 0)$ and $(0, \infty)$

3. 1

ANSWERS TO SECTION QUIZ

1.   The domain of $\sin^{-1}x$ is only $[-1, 1]$,   so $\sin^{-1}(\sin x) = x$ only if

x is in $[-\pi/2, \pi/2]$.

2.   (a) $\sqrt{1 - x^2}/x$

(b) $4x/(4 + x^2)$

(c) $\pi/4$

3. $2 + x^2 > 1$ and the domain of $\sin^{-1}x$ is $[-1, 1]$.

4.   It is θ such that $\tan \theta = \pi/2$,   i.e., $\theta \approx 1.00$.

5.   (a) $2/\sqrt{1 - (2x_0 - 2)^2}$

(b) $[1/2, 3/2]$

(c) $\sin(\sin^{-1}(2x - 2)) = 2x - 2$; $\cos(\sin^{-1}(2x - 2)) = \sqrt{1 - (2x - 2)^2}$;

$\tan(\sin^{-1}(2x - 2)) = (2x - 2)/\sqrt{1 - (2x - 2)^2}$.

5.5   Graphing and Word Problems

PREREQUISITES

1.   Recall how to solve minimum-maximum word problems (Section 3.5).

2.   Recall how to use derivatives to aid in graphing (Section 3.4).

3.   Recall how to differentiate trigonometric functions (Section 5.2).

4.   Recall how to use the chain rule for differentiation (Section 2.2).

5.   Recall the concept of periodicity (Section 5.1).

PREREQUISITE QUIZ

1.   Maximize the area of a rectangle whose perimeter is 8 cm.

2.   Differentiate:

   (a) $\sin^2 x$

   (b) $\sin x/\sqrt{\cos x}$

   (c) $\cot x$

3.   If $f(x) = \sin(4\pi x + 2)$, what is the minimum period?

4.   Explain how the first derivative can be used as an aid in graphing.

5.   What information is given by the sign of the second derivative?

GOALS

1.   Be able to solve word problems involving angles.

2.   Be able to graph functions involving trigonometric functions.

STUDY HINTS

1.   Typical word problem. Example 1 should be studied carefully. The word
   "revolution" should immediately translate into "2π radians" in your head.
   Also, since angles are involved, try to relate given and unknown quantities
   in terms of trigonometric functions.

2. <u>Review and practice</u>. Many of these problems rely on the chain rule to

    relate distances, angles, and time. If necessary, review the chain rule

    and problem solving techniques from Section 3.5. As always, practice is

    important.

3. <u>Graphing</u>. The usual techniques of graphing carry over to trigonometric

    functions. One new aspect is periodicity, which simplifies your work

    tremendously. Note any symmetry or periodicity first. Note that if a

    function is a sum of trigonometric functions, the period is the largest

    of the periods of the terms.

SOLUTIONS TO EVERY OTHER ODD EXERCISE

1.

From the diagram, note that $\tan \theta = h(t)/200 =$
$(1000 - 40t - 16t^2)/200$. Consider θ as a
function of t and differentiate both sides
with respect to t . $\sec^2\theta \ (d\theta/dt) = -(40/200) -$
$(16/200) 2t$. At $t = 4$, $\tan \theta = [1000 - 40(4)$

$16(4)] /200 = 73/25$; hence $\sec^2\theta\big|_{t=4} = 1 + (73/25)^2 = 9.5264$ because
$\sec^2\theta = \tan^2\theta + 1$. Plug in $\sec^2\theta\big|_{t=4}$ and $t = 4$, and solve for
$d\theta/dt$. We get $d\theta/dt\big|_{t=4} = -0.088$ rad/sec. The negative sign indicates
that the angle is decreasing.

5.

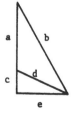

We know that $a + c = 10$ and $a + b = 20$. Differentiating both sides with respect to time, we
get $a' + c' = 0$ and $a' + b' = 0$. Therefore,
$-a' = b' = c'$. By the Pythagorean theorem, we
have $(a + c)^2 + e^2 = b^2$ and $c^2 + e^2 = d^2$.

Differentiating $(a + c)^2 + e^2 = b^2$ gives $2(a + c)(a' + c') + 2ee' =$
$2bb'$ or $2e/b = b'$. Differentiation of $c^2 + e^2 = d^2$ gives

5. (continued)

$2cc' + 2ee' = 2dd'$ or $(cc' + ee')/d = d'$, where $c' = b' = 2e/b$ and e' is given as 2. After 3 seconds, $e = 6$; $b = \sqrt{(10)^2 + (6)^2} = \sqrt{136} \approx 11.66$; $a = 20 - \sqrt{136} \approx 8.34$; $c = 10 - (20 - \sqrt{136}) \approx 1.66$; $d = \sqrt{c^2 + (6)^2} \approx 6.23$; $c' = 1.03$. Therefore, $d' \approx 2.20$ meters/second.

9. We will maximize the square of the distance; $d^2 = (3 \sin 3t - 3 \cos 2t)^2 + (3 \cos 3t - 3 \sin 2t)^2$. Both particles are moving in the same circle of radius 3, but at different speeds; therefore, the maximum distance should be no more. than 6. $(d^2)' = 2(3 \sin 3t - 3 \cos 2t)(9 \cos 3t + 6 \sin 2t) + 2(3 \cos 3t - 3 \sin 2t)(-9 \sin 3t - 6 \cos 2t) = 2[-45 \cos 2t \times \cos 3t + 45 \sin 2t \sin 3t] = -90 \cos 5t$. Thus, the critical points are at $(2n + 1)\pi/10$, where n is an integer. At $\pi/10$, $d = 0$, but at $3\pi/10$, $d = 6$, so the maximum distance between A and B is 6. (Notice that if we maximize the distance itself, the derivative is more complicative; however, the critical points are found by solving essentially the same equation.)

13.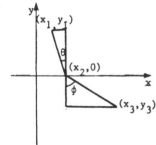

The total time is $T = \sqrt{(x_2 - x_1)^2 + y_1^2} + \sqrt{(x_3 - x_2)^2 + y_3^2}/3$. To minimize the time, we find x_2 such that $dT/dx_2 = 0$. $T' = (1/2) \times [(x_2 - x_1)^2 + y_1^2]^{-1/2}2(x_2 - x_1) + (1/3)(1/2) \times [(x_3 - x_2)^2 + y_3^2]^{-1/2}2(x_3 - x_2)(-1)$. Thus, we must solve $(x_2 - x_1)/\sqrt{(x_2 - x_1)^2 + y_1^2} = (1/3)(x_3 - x_2)/\sqrt{(x_3 - x_2)^2 + y_3^2}$. As in

Example 2, the left-hand side is $\sin \theta$ and the right-hand side is $(1/3) \sin \phi$. Therefore, we have $3 \sin \theta = \sin \phi$.

17. $f'(x) = \sin x + x \cos x - 2 \sin x = x \cos x - \sin x$; $f''(x) = \cos x -$

$x \sin x - \cos x = -x \sin x$. $f''(x) = 0$ at $x = \pm n\pi$, $n = a$ non-

negative integer. When   x   is positive, $f''(x) < 0$ for $(0,\pi)$,

$(2\pi,3\pi)$,   ... , $(2n\pi,(2n + 1)\pi)$.   When   x   is negative, $f''(x) < 0$

for $(-\pi,0)$, $(-3\pi,-2\pi)$,   ... , $(-(2n + 1)\pi,-2n\pi)$.   Thus, $f(x)$

is concave downward in $(2n\pi,(2n + 1)\pi)$ for $x > 0$ and in

$(-(2n + 1)\pi,-2n\pi)$ for $x < 0$. $f(x)$ is concave upward everywhere else.

21.

$f'(x) = 1 - \sin x$,   so the critical points
occur at $x = \pi/2 + 2n\pi$,   where   n   is an
integer. $f''(x) = -\cos x$,   so possible inflection points occur at $x = \pi/2 + n\pi$.
Note that the graph oscillates around the
line $y = x$.

25. $x^{2/3}\cos x$ is symmetric about the y-axis. Its graph oscillates between
$\pm x^{2/3}$.   Consider $f(x) = x^{2/3}$: $f'(x) = 2/3x^{1/3}$ and $f''(x) = -2/9x^{4/3}$.
There is a cusp at $x = 0$ and the graph of $x^{2/3}$ is concave down. The
graph of $x^{2/3}\cos x$ looks like $\cos x$ except for the scaling factor.

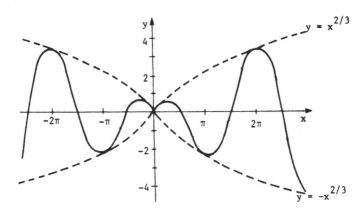

29. Recall that the equation of a line is $y = y_1 + m(x - x_1)$, where m
    is the slope. $f'(x) = -2/\sqrt{1 - 4x^2} - 1/\sqrt{1 - x^2}$, which is -3 at
    $x = 0$. Therefore, the slope of the tangent line is -3 and the slope
    of the normal line is the negative reciprocal, $1/3$. Thus, the tangent
    line is $y = -3x + \pi$ and the normal line is $y = x/3 + \pi$.

33. $f'(x) = (x \cos x - \sin x)/x^2$. The critical points occur where
    $x \cos x = \sin x$ or $x = \tan x$, which has infinitely many solutions.
    As $x \to \infty$, $\tan x$ also approaches ∞ . This also occurs near $\pi/2 + n\pi$,
    where n is an integer. Thus, the local maxima and minima are about
    π units apart as $x \to \infty$.

SECTION QUIZ

1. Sketch $y = \sqrt{x} \cos x + \cos \pi$.

2. You and your partner have just finished cutting down a 100 m. redwood
    tree. As you yell "Timberr...", your partner realizes that the tree is
    about to fall on top of him. Without thinking, he makes a mad dash in
    the path of the tree's shadow. If the tree is falling at a rate of
    $9°$/sec., how fast is it falling vertically after falling an angle of
    $\pi/4$? How fast is the shadow increasing assuming the sun is directly
    overhead? (Your friend is a world class sprinter and barely escaped
    being crushed.)

3. Those silly sisters, Sandra and Sheila, are at it again. On a pleasant
    Spring day, they decided to go ballooning. As Sandra unties the rope
    anchoring the balloon, a sudden gust of wind whisks the balloon away,
    parallel to the ground with Sheila in it at a speed of 1 m./sec. In
    the meantime, Sandra, who has been hanging on for dear life, falls to

3.   (continued)

the ground, 25 meters below the balloon. She gets up and runs after Sheila at a speed of 0.25 m/sec. After one minute, Sandra tires and yells, "Bye, Sheila. I'll send the Air Force to rescue you." How fast is the angle of Sandra's eyes changing one minute after she quit running if she continues to keep an eye on the balloon?

ANSWERS TO PREREQUISITE QUIZ

1. 4 cm^2 (2 cm. × 2 cm.)

2.   (a) $2 \sin x \cos x$

(b) $\sqrt{\cos x} + \sin^2 x / 2 \cos^{3/2} x$

(c) $- \csc^2 x$

3. $1/4$

4.   If $f'(x) < 0$, the function is decreasing. If $f'(x) > 0$, the function is increasing.

5.   If $f''(x) < 0$, the function is concave downward. If $f''(x) > 0$, the function is concave upward.

ANSWERS TO SECTION QUIZ

1.

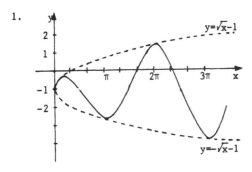

2. $5\pi/\sqrt{2}$ m./sec. ; $5\pi/\sqrt{2}$ m./sec.

3. $-1/2650$ rad./sec.

5.6   Graphing in Polar Coordinates

PREREQUISITES

1.   Recall the first derivative test (Section 3.2).

2.   Recall how to calculate with polar coordinates (Section 5.1).

PREREQUISITE QUIZ

1.   Convert the cartesian coordinates   (4,2)   to polar coordinates.

2.   Convert the polar coordinates   (-3,-3π/4)   to cartesian coordinates.

3.   What do you know about a function   f   if $f'(x_0) > 0$?

GOALS

1.   Be able to graph a function given in polar coordinates on the xy-plane.

2.   Be able to find the slope of a function given in polar coordinates.

STUDY HINTS

1.   Symmetry. As with graphing of functions in the xy-plane, symmetry can
     be used to eliminate much work. Example 1 shows how to determine symmetry
     in the   x- and y-axes.

2.   Rotational symmetry. Note that if $f(\theta + \phi) = f(\theta)$ for all   θ , then
     the graph is unchanged after a rotation of   φ   radians.

3.   Cartesian vs. polar coordinates graphing. The graph of $r = f(\theta)$ can be
     graphed in two ways. Graphing in the   rθ-plane was discussed in Section
     5.1 and graphing in the xy-plane is discussed in this section. Be care-
     ful to read what is asked for.

4.   Rose petals. Examples 1 and 2 illustrate a general fact about   r=sin nθ
     and   r=cos nθ . If n is odd, the graph is an n-petaled rose. If   n   is
     even, the graph is a 2n-petaled rose.

5. <u>Tangents in polar coordinates</u>. Rather than memorizing the formula

dy/dx = [(tan θ)dr/dθ + r] / [(dr/dθ) - r tan θ] ,   you may find it

easier to derive the formula. Just remember that dy/dx = (dy/dθ)/

(dx/dθ) and use the chain rule to differentiate x = r cos θ and

y = r sin θ .

6. <u>Interpretation of maxima, minima</u>. Sometimes, a minimum represents a

<u>maximal</u> distance from the origin because it is usually a negative num-

ber and distances are absolute values. A minimum represents an actual

minimal distance if it is positive. What are the corresponding state-

ments for maxima?

SOLUTIONS TO EVERY OTHER ODD EXERCISE

1. 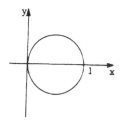 In cartesian coordinates, we have $r = \sqrt{x^2 + y^2} =$

$x/\sqrt{x^2 + y^2} = \cos\theta$ or $x^2 + y^2 = x$. Rearrange-

ment yields $(x^2 - x + 1/4) + y^2 = (x - 1/2)^2 +$

$y^2 = 1/4$. So the graph is a circle centered at

(1/2,0) with radius 1/2 .

5. 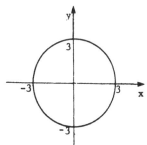 $r = 3$ implies $r^2 = 9$ or $x^2 + y^2 = 9$.

This is a circle of radius 3 centered at

the origin.

9.

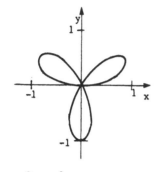

$r = \sin 3\theta = \sin(2\theta + \theta) = \sin 2\theta \cos \theta +$
$\cos 2\theta \sin \theta = (2 \sin \theta \cos \theta)\cos \theta +$
$(\cos^2\theta - \sin^2\theta) \sin \theta = 3 \sin \theta \cos^2\theta -$
$\sin^3\theta$. Multiply both sides by r^3 to
get $r^4 = 3r^3 \sin \theta \cos^2\theta - r^3 \sin^3\theta$.
Substituting $r^2 = x^2 + y^2$, $x = r \cos \theta$,
and $y = r \sin \theta$, we get $(x^2 + y^2)^2 =$

$3x^2y - y^3$. The graph of $\sin 3\theta$ is exactly like $\cos 3\theta$ except that
$\sin 3\theta$ is shifted $(\pi/2)/3 = \pi/6$ radians to the right. The graph of
$r = \sin 3\theta$ is obtained by rotating $r = \cos 3\theta$ (Example 2) by $\pi/6$.

13. $f(\pi/2 - \theta) = f(\theta)$ implies that $f(\theta)$ is symmetric with respect to the
diagonal line $\theta = \pi/4$. Hence, the graph of $r = f(\theta)$ is symmetric
with respect to the line $\theta = \pi/4$.

17. Use the relations $x = r \cos \theta$ and $y = r \sin \theta$ to get $1 = x^2 + xy +$
$y^2 = r^2 \cos^2\theta + (r \cos \theta)(r \sin \theta) + r^2 \sin^2\theta = r^2(\cos^2\theta + \sin^2\theta) +$
$r^2 \cos \theta \sin \theta = r^2(1 + \cos \theta \sin \theta)$.

21. Use the relations $x = r \cos \theta$ and $y = r \sin \theta$ to get $y = r \sin \theta =$
$r \cos \theta + 1 = x + 1$. This simplifies to $r(\sin \theta - \cos \theta) = 1$.

25. The slope of the tangent line to the graph $r = f(\theta)$ at (r,θ) is
$[(\tan \theta)(dr/d\theta) + r]/[dr/d\theta - r \tan \theta]$. Here, $dr/d\theta = 10 \cos 5\theta$.
At $\theta = \pi/2$, $r = 2 \sin 5\theta = 2$ and $dr/d\theta = 0$, so the slope is
$[(\tan \theta)(0) + 2]/[0 - 2 \tan \theta]$. Since $\tan \theta \to \infty$, we need to find
the limit. Divide by $\tan \theta/\tan \theta$ to get $(0 + 2/\tan \theta)/(-2)$, which
has a limit of 0 .

29. The slope of the tangent line to the graph $r = f(\theta)$ at (r,θ) is

$[(\tan \theta)(dr/d\theta) + r] / [dr/d\theta - r \tan \theta]$. Here, $dr/d\theta = 3 \cos \theta -$

$2\theta \sin(\theta^2)$. At $\theta = 0$, $r = 1$, $\tan \theta = 0$, and $dr/d\theta = 3$. Thus,

the slope is $[0(3) + 1]/[3 - (1)(0)] = 1/3$.

33.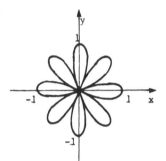

Note that $dr/d\theta = -4 \sin 4\theta$, which is

0 if $\theta = n\pi/4$ where n is an integer.

$d^2r/d\theta^2 = -16\cos 4\theta$ which is negative for

even n and positive for odd n . Hence

$r = 1$ at $\theta = n\pi/2$, which are the local

maxima and $r = -1$ at $\theta = (2n + 1)\pi/4$,

which are the local minima.

37.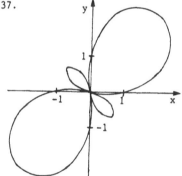

Note that $dr/d\theta = 4 \cos 2\theta$, which is 0

if $2\theta = (2n + 1)\pi/2$, i.e., $\theta = (2n + 1)\pi/$

4 for all integers n . Also, $d^2r/d\theta^2 =$

$-8 \sin 2\theta$. For n even, $\theta = (2n + 1)\pi/4$

is a local maximum point with $r = 3$; for

n odd, $\theta = (2n + 1)\pi/4$ is a local minimum

point with $r = -1$.

SECTION QUIZ

1. How is the graph of $\cos 2(\theta + 7\pi/8)$ related to the graph of $\cos 2\theta$?

2. If $r = \csc \theta$, what is the equation of the tangent line at $\theta = \pi/4$?

3. Sketch the graph of $\sin^{-1}r = 2\theta$. Are there any restrictions on r ?

4. In a hospital waiting room, an expectant father was seen pacing and

    making a track in the carpet. Another gentleman, waiting for his fourth

    child, notices that the track can be described by $r = \sin(\theta + \pi/6)$.

4.   (a)   Sketch the curve.

    (b)   When the nurse enters and says, "Congratulations, they're triplets,"

         our new father runs off to buy cigars. His path is a tangent from

         the point $(r,\theta) = (1/2,4\pi)$. What equation describes the path?

ANSWERS TO PREREQUISITE QUIZ

1. $(\sqrt{20},\tan^{-1}0.5)$

2. $(3/\sqrt{2},3/\sqrt{2})$

3.   f is increasing at x_0 .

ANSWERS TO SECTION QUIZ

1.   The graph of $\cos 2(\theta + 7\pi/8)$ is the same as the graph of $\cos 2\theta$ after

    a clockwise rotation of $7\pi/8$ radians.

2. $y = 1$

3.

The restriction is $-1 \leqslant r \leqslant 1$. The graph

is that of $r = \sin 2\theta$.

4.   (a)

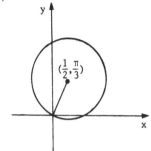

$(\frac{1}{2}, \frac{\pi}{3})$

(b) $y = (x - 1/2)/\sqrt{3}$

5.S   Supplement to Chapter 5:   Length of Days

SOLUTIONS TO EVERY OTHER ODD EXERCISE

1.   Using the first-order approximation, $\Delta S \approx dS/dT$ with dS/dT given by

formula (4) and $\ell = 33.57°$, $\alpha = 0.41$ radian and $T = 22$ (June 21

is day 1). Note that $2\pi T/365$ is in radians. We get $dS/dT|_{T=22} =$

-0.0083 hours ≈ -30 seconds. Hence on July 13, the sun sets at about

8:05:30 . On July 14, we can use the same formula for dS/dT with $T = 23$

We get $dS/dT|_{T=23} = -0.0086$ hour ≈ -31 seconds. Hence on July 14, the

sunsets at about 8:05:00 .

5.   We use the formula: $\sin A = \cos\ell\sqrt{1 - \sin^2\alpha\cos^2(2\pi T/365)}\cos(2\pi t/24) +$

$\sin\ell\sin\alpha\cos(2\pi T/365)$. At sunset, $A = 0$ and $t = 6.5$, so we get

$\cos\ell\sqrt{1 - \sin^2\alpha\cos^2(2\pi T/365)}\cos(13\pi/24)$ $-\sin\ell\sin\alpha\cos(2\pi T/365)$,

which implies $\sqrt{1 - \sin^2\alpha\cos^2(2\pi T/365)}/\cos(2\pi T/365) = -\tan\ell\sin\alpha/$

$\cos(13\pi/24)$. Squaring gives us $1/\cos^2(2\pi T/365) - \sin^2\alpha = [\tan\ell\sin\alpha/$

$\cos(13\pi/24)]^2$. Adding $\sin^2\alpha$ to both sides, taking the square root,

the reciprocal, and then arccos gives us $2\pi T/365 = \cos^{-1}[\tan^2\ell\sin^2\alpha/$

$\cos(13\pi/24) + \sin^2\alpha]^{-1/2}$. Multiplying through by $365/2\pi$ gives us

$T = (365/2\pi)\cos^{-1}\sqrt{\cos^2(13\pi/24)/[\sin^2\alpha(\tan^2\ell + \cos^2(13\pi/24))]} =$

$(365/2\pi)\cos^{-1}\sqrt{0.107/(\tan^2\ell + 0.017)}$. The domain requires the expression

under the radical to be in $[0,1]$, so we solve $0 \leqslant \cos^2(13\pi/24)/$

$[\sin^2\alpha(\tan^2\ell + \cos^2(13\pi/24))] \leqslant 1$. Divide by $\cos^2(13\pi/24)/\sin^2\alpha$ and

take the reciprocal to get $\infty > \tan^2\ell + \cos^2(13\pi/24) \geqslant \cos^2(13\pi/24)/\sin^2\alpha$.

Subtract $\cos^2(13\pi/24)$, take the square root and then the arctangent.

Thus, the domain is $[-\pi/2, -0.29]$ and $[0.29, \pi/2]$. We plot the following

points:

5.   (continued)

$\ell(°)$	16.71	17	18	20	25	30	40	50
±T (days)	0.0	10.2	21.1	32.6	48.1	57.2	68.3	75.2

$\ell(°)$	60	70	80	90
±T (days)	80.2	84.3	87.9	91.3

The other half of the graph is obtained

by a reflection across the T-axis.

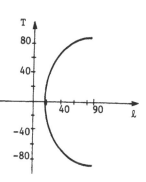

9.   (a)   We use the formula for the length of days. For this problem, we
have $\ell = 0$ because the location is at the equator;   T = 0   by
definition; and   dA/dt   is the quantity of interest. Therefore,
the equation reduces to $\sin A = \sqrt{1 - \sin^2\alpha} \, [\cos(2\pi t/24)]$. Differentiating both sides with respect to   t   gives   cos A(dA/dt) =
$\sqrt{1 - \sin^2\alpha} \, [-\sin(2\pi t/24)] (2\pi/24)$. The formula tells us that all
days at the equator are 12 hours long, so   t = 18   at sunrise.
Also, we know that   A = 0   at sunrise, so $(dA/dt) = (\pi/12) \times$
$\sqrt{1 - \sin^2\alpha} = 0.24$ radians/hour or   13.76°/hour .

(b)   The linear approximation tells us that $\Delta A = (dA/dt)\Delta t$,   so
$\Delta t = \Delta A/(dA/dt) = 5°/(13.76°/hr.) = 0.36$ hours = 21 minutes, 49
seconds.

5.R Review Exercises for Chapter 5

SOLUTIONS TO EVERY OTHER ODD EXERCISE

1. Since π radian = $180°$, we get $66° \times (\pi/180) = 1.152$ radians.

5. Use $x = r \cos \theta$ and $y = r \sin \theta$ to get $y = r \sin \theta = r^2 \cos^2 \theta = x^2$

or $\tan \theta \sec \theta = r$.

9. By the definition of sine, we have $\sin 8° = a/10$, so $a = 10 \sin 8° \approx$

1.392 .

13. 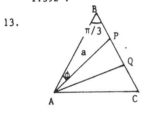 By the law of cosines, $a^2 = 1^2 + 3^2 -$

$2(1)(3)\cos(\pi/3) = 7$ so $a = \sqrt{7}$. Applying

the law of cosines again, $1^2 = 3^2 + (\sqrt{7})^2 -$

$2(3)(\sqrt{7})\cos \phi$ implies $\cos \phi = 15/6\sqrt{7}$, so

$\phi \approx 0.333$. By symmetry, the lower angle is also 0.333 and since all

of the angles of an equilateral triangle are $\pi/3$, the remaining angle

is 0.380 .

17. Since $(d/dx)\sin x = \cos x$, we get $dy/dx = 1 + \sin 3x + 3x \cos 3x$.

21. Using the fact that $(d/dy) \tan y = \sec^2 y$, we get $h'(y) = 3y^2 +$

$2 \tan(y^3) + 2y \sec^2(y^3) \cdot (3y^2) = 3y^2 + 2 \tan(y^3) + 6y^3 \sec^2(y^3)$.

25. Using the facts that $(d/dx)\sec^{-1}x = 1/x\sqrt{x^2 - 1}$ and $(d/dx)\sin x =$

$\cos x$, we get $f'(x) = \{1/[(x + \sin x)^2\sqrt{(x + \sin x)^4 - 1}]\}\cdot 2(x + \sin x) \times$

$(1 + \cos x) = 2(1 + \cos x)/(x + \sin x)\sqrt{(x + \sin x)^4 - 1}$.

29. Since $(d/dx)\sin^{-1}x = 1/\sqrt{1 - x^2}$, we get $(1/\sqrt{1 - x})(d/dx)\sqrt{x} =$

$1/2\sqrt{x}\sqrt{1 - x} = 1/2\sqrt{x - x^2}$.

33. Using the fact that $(d/dx)\sin^{-1}x = 1/\sqrt{1 - x^2}$ and $(d/dx)\cos x = - \sin x$,

we get $(d/dx)\sin^{-1}(\sqrt{x} + \cos 3x) = [1/\sqrt{1 - (\sqrt{x}\cos 3x)^2}]\cdot(1/2\sqrt{x} - 3 \sin 3x)$

37. By the product rule, $dh/dx = \sin^{-1}(x + 1) + x/\sqrt{1 - (x + 1)^2}$. By the chain rule, $dh/dy = (dh/dx)(dx/dy) = \sin^{-1}(x + 1) + x/\sqrt{1 - (x + 1)^2}] \times (1 - 3y^2) = \sin^{-1}(y - y^3 + 1) + (y - y^3)/\sqrt{1 - (y - y^3 + 1)^2}](1 - 3y^2)$.

41. Guess that $\int \sin 3x\, dx = a \cos 3x + C$. Differentiation yields $a = -1/3$; therefore, $\int \sin 3x\, dx = -\cos 3x/3 + C$.

45. Guess that $\int (3x^2 \sin x^3)dx$ has the form $a \cos x^3 + C$. Differentiation shows that $a = -1$, so the integral is $-\cos x^3 + x^2 + C$.

49. Factor out $1/4$, so $(1/4)\int [dy/(1 + (y/2)^2)]$ should have the form $a \tan^{-1}(y/2)$. Differentiation shows that $a = 1/2$; therefore, $\int [1/(4 + y^2)]\, dy = (1/2) \tan^{-1}(y/2) + C$.

53. Factor out $1/2$ to get $\int_{-1}^{1} [1/\sqrt{4 - x^2}]\, dx = (1/2)\int_{-1}^{1} [1/\sqrt{1 - (x/2)^2}]\, dx$. The antiderivative should have the form $a \sin^{-1}(x/2) + C$. Differentiation yields $(a/2)/\sqrt{1 - (x/2)^2}$, so $a = 1$ and the integral is $\sin^{-1}(x/2)\big|_{-1}^{1} = \pi/6 - (-\pi/6) = \pi/3$.

57. (a) We want to find an interval in which $f(x)$ is strictly increasing or decreasing. $f'(x) = 3x^2 - 3$ implies the critical points are ± 1; therefore, the interval containing zero is $[-1,1]$.

    (b) $f'(x) < 0$ in $[-1,1]$, so it is decreasing and the domain of g is $[f(b), f(a)] = [5,9]$.

    (c) By inspection, $f(0) = 7$, so $g(7) = 0$. Therefore, by the inverse function rule, $g'(7) = 1/[f'(0)] = 1/(3x^2 - 3)\big|_0 = -1/3$.

61.

By similar triangles, $30/(10 + x) = y/x$, which implies $30x = y(10 + x)$. Upon differentiating with respect to t, we have $30x' = y'(10 + x) + yx'$, and so $x' = y'(10 + x)/(30 - y)$. After 4 seconds, $y = 8$, $x = 40/11$, and $y' = 2$. Therefore, $x' = 2(10 + 40/11)/(30 - 8) = 2(150/11)/22 = (150/121)$ m./sec.

65.   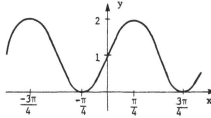 Time is distance/speed, so we want to minimize $T = \sqrt{1 + x^2}/4 + (4 - x)/16$. $T'(x) =$
$(1/2)(1 + x^2)^{-1/2}(2x)/4 - 1/16 = (4x -$
$\sqrt{1 + x^2})/16\sqrt{1 + x^2}$. Setting $T'(x) = 0$,

we get $4x = \sqrt{1 + x^2}$ or $16x^2 = 1 + x^2$; therefore, $x = (1/\sqrt{15})$ km is
the point to where she should row.

69.   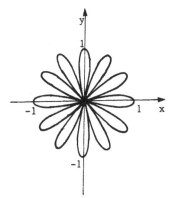 $f'(x) = 2[(\cos x)(\cos x) +$
$(\sin x)(-\sin x)] = 2(\cos^2 x -$
$\sin^2 x) = 2 \cos 2x$, so the
critical points are $(2n + 1)\pi/4$,
where n is an integer. $f''(x) =$
$-4 \sin 2x$, so the inflection points are $n\pi/2$. $f''(x) < 0$ in $[n\pi,$
$(2n + 1)\pi/2]$, so $f(x)$ is concave down in this interval and concave
up elsewhere. Since $\sin(x + \pi) = -\sin x$ and $\cos(x + \pi) = -\cos x$, we
have $f(x + \pi) = f(x)$. Therefore, the period is π and we only have to
graph $0 \leqslant x \leqslant \pi$ and then use the symmetry properties.

73.   We note that $f(\theta) = f(-\theta)$;
$f(\pi - \theta) = \cos(6\pi - 6\theta) = \cos(-6\theta) =$
$\cos 6\theta = f(\theta)$; $f(\pi + \theta) =$
$\cos(6\pi + 6\theta) = \cos(6\theta) = f(\theta)$. Thus,
the graph is symmetric in the x-axis,
the y-axis, and the origin, so we need
only plot the graph in $[0, \pi/2]$ and
use reflections. $dr/d\theta = -6 \sin 6\theta$
implies that critical points occur at 0 , $\pi/6$, $\pi/3$, $\pi/2$. The zeros
occur at $\pi/12$, $\pi/4$, $5\pi/12$.

77. The tangent line is given by $y = y_0 + m(x - x_0)$, where $y_0 = r \sin \theta$,

$x_0 = r \cos \theta$, and $m = [(\tan \theta)(dr/d\theta) + r]/[dr/d\theta - r \tan \theta]$. Here

$dr/d\theta = -4 \sin 4\theta$. At $\theta = \pi/4$, $\tan(\pi/4) = 1$, $r = \cos \pi = -1$,

$x_0 = (-1) \cos (\pi/4) = -\sqrt{2}/2$, and $y_0 = (-1) \sin(\pi/4) = -\sqrt{2}/2$. Thus,

$m = [(1)(0) + (-1)]/[0 - (-1)(1)] = -1$, and the tangent line is $y = $

$(-\sqrt{2}/2) + (-1)[x - (-\sqrt{2}/2)] = -x - \sqrt{2}$.

81. We use the chain rule and the inverse function rule. Let $y = \sqrt{x}$,

then the chain rule tells us that $g'(x) = (dg/dy) \cdot (dy/dx)$. Now

according to the inverse function rule, $dg/dy = 1/f'(g(y)) = 1/$

$f'(f^{-1}(y)) = 1/f'(f^{-1}(\sqrt{x}))$, and by ordinary differentiation, $dy/dx =$

$1/2\sqrt{x}$. Therefore, $g'(x) = 1/2f'(f^{-1}(\sqrt{x}))\sqrt{x}$.

85. If $x \neq 0$, then $f(x) = x^4 \sin(1/x)$ and $f'(x) = 4x^3 \sin(1/x) +$

$x^4(-1/x^2)\cos(1/x) = 4x^3 \sin(1/x) - x^2 \cos(1/x)$. Therefore, the chain

rule and the product rule may be applied again to get $f''(x) =$

$12x^2 \sin(1/x) + 4x^3(-1/x^2) \cos(1/x) - 2x \cos(1/x) - x^2(-1/x^2)(-\sin(1/x)) =$

$12x^2 \sin(1/x) - 6x \cos(1/x) - \sin(1/x)$. Since the last term has no limit

as $x \to 0$, $f''(x)$ has no limit as $x \to 0$.

   On the other hand, $f'(0) = \lim\limits_{x \to 0}\{[f(x) - f(0)]/x\} = \lim\limits_{x \to 0}\{[x^4 \sin(1/x)]/$

$x\} = \lim\limits_{x \to 0}[x^3 \sin(1/x)]$. This limit is zero since $|\sin(1/x)| \leq 1$. Also,

$f''(0) = \lim\limits_{x \to 0}\{[f'(x) - f'(0)]/x\} = \lim\limits_{x \to 0}\{[4x^3 \sin(1/x) - x^2 \cos(1/x)]/x\} =$

$\lim\limits_{x \to 0}[4x^2 \sin(1/x) - x \cos(1/x)]$. Again, the limit is 0 since

$|\sin(1/x)| \leq 1$ and $|\cos(1/x)| \leq 1$. Thus, f is indeed twice differ-

entiable, but f'' is not continuous since $\lim\limits_{x \to 0} f''(x)$ does not exist.

TEST FOR CHAPTER 5

1. True or false.

    (a) The origin can have any θ in polar coordinates.

    (b) In polar coordinates, $\sin \theta = 1$ and $\sin \theta = -1$ describe the
         same line.

    (c) $(d/dx) \sin^{-1}x = -\cos x/\sin^2 x$.

    (d) $y = x^5$ has an inverse on $(-\infty, \infty)$, but $y = x^4$ does not.

    (e) If $f'(x) < 0$ on $[a,b]$, then the domain of f^{-1} is
         $[f(a), f(b)]$.

2. Differentiate:

    (a) $(\cos \sqrt{x} \sin x)^2$

    (b) $\sec^{-1}(2x)$

    (c) $\tan^{-1}(5x + 2)$

    (d) $f^{-1}(x)$ if $f(x) = x + (1/2) \cos x$

3. Integrate:

    (a) $\int_0^\pi \sin t \, dt$

    (b) $\int_0^1 [d\theta/(1 + \theta^2)]$

    (c) $\int \csc x \sin^2 x \, dx$

4. (a) Sketch the graph of $y = \cos 2(x + 3\pi/8)$.

    (b) Find intervals for which an inverse exists.

    (c) Sketch the inverse on the interval which contains $x = \pi/4$.

5. (a) Sketch the graph of $r = \cos 2(\theta + 3\pi/8)$ in the xy-plane. (Com-
         pare this sketch with that of Question 4(a).)

    (b) Convert this equation to cartesian coordinates.

6. In each case, find $\sin 2\theta$ for the given information.

    (a) $\sin \theta = 5/6$

    (b) $\tan^{-1}x = \theta$

7. A particle follows the path given by the equation $r = 2 \cos \theta + \sin \theta$.

    (a) If it flies off along the tangent line at $\theta = \pi/4$, what is the equation of the tangent line?

    (b) Will the particle hit the x-axis, y-axis, both, or neither, assuming the particle traverses a counterclockwise path? Where will it hit?

8. Let $x = \pi/4 + 0.2$. Approximate $\cos x$, $\sin x$, and $\tan x$. Use $\sqrt{2} \approx 1.4$.

9. Let $x = 2t$ and $y = \cos^{-1}t$.

    (a) What is the domain?

    (b) Sketch the curve in the xy-plane.

10. Steve, the fearless skydiver, was attempting to dive into a red target. 250 m. above the target, he was falling at 10 m./sec. Unfortunately, skydiving Steve's friends placed the target in the middle of a bull pasture. One angry bull stood 50 m. from the target. How fast is the bull's head rotating downward as he keeps an eye on Steve?

ANSWERS TO CHAPTER TEST

1. (a) True

    (b) True

    (c) False; $(d/dx) \sin^{-1}x = 1/\sqrt{1 - x^2}$.

    (d) True

    (e) False; if $f'(x) < 0$, the domain is $[f(b),f(a)]$.

2. (a) $2(\cos \sqrt{x} \sin x)(-\sin \sqrt{x} \sin x/2\sqrt{x} + \cos \sqrt{x} \cos x)$

   (b) $1/x\sqrt{4x^2 - 1}$

   (c) $5/[(5x + 2)^2 + 1]$

   (d) $1/(1 - (1/2)\sin x)$

3. (a) 2

   (b) $\pi/4$

   (c) $-\cos x + C$

4. (a)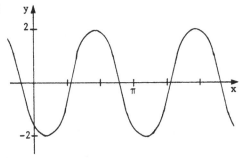

   (b) $(n\pi/2 , (n + 1)\pi/2)$, where n is an integer.

   (c)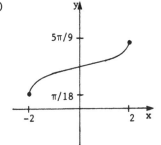

5.   (a)

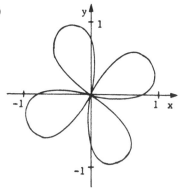

(b) $\sqrt{Z} = (y^2 - 2xy - x^2)/(x^2 + y^2)^{3/2}$

6.   (a) $5\sqrt{\text{TT}}/18$

(b) $2x/(1 + x^2)$

7.   (a) $2y + x = 9/2$

(b)   y-axis at   (0,9/4)

8. $\cos (\pi/4 + 0.2) \approx 0.56$; $\sin (\pi/4 + 0.2) \approx 0.84$; $\tan (\pi/4 + 0.2) \approx 1.40$

9.   (a) $-1 \leqslant t \leqslant 1$ or $-2 \leqslant x \leqslant 2$

(b)

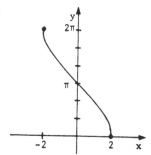

10.   (1/130) rad/sec

6.1 Exponential Functions

PREREQUISITES

1. Recall the laws of exponents for rational powers (Section R.3).

PREREQUISITE QUIZ

1. Simplify the following without using a calculator:

   (a) $6^{25}/6^{23}$

   (b) $(8^3 \cdot 2^3)^{1/2}$

   (c) $(1/8)^{-1/3}$

GOALS

1. Be able to manipulate exponential functions.

2. Be able to recognize the graphs of exponential functions.

STUDY HINTS

1. Exponential growth. In general, if something grows by a factor of b per unit of time, the growth factor after t units of time is b^t .

2. Real powers. The properties are an extension from rational to real powers. The same properties that held for rational powers now hold for real powers.

3. <u>Notation.</u> $\exp_b x$ is the same as b^x .

4. <u>Exponential graphs.</u> You should know the general shapes of the graphs. If $b > 1$, then b^x starts near 0 for $x \to -\infty$, increases until it passes through $(0,1)$, and continues to increase very steeply toward ∞ as $x \to \infty$. Note that if $0 < b < 1$, then $b^x = (1/b)^{-x}$; therefore, $(1/b)^x$ is a reflection of b^x across the y-axis.

SOLUTIONS TO EVERY OTHER ODD EXERCISE

1. (a) In 4 hours, the culture triples its mass twice, so it grows by a factor of $3 \cdot 3 = 9$.

   (b) In 6 hours, the mass triples three times or by a factor of $3^3 = 27$.

   (c) If it grows by a factor of k in 1 hour, it grows by a factor of $k \cdot k = k^2$ in 2 hours. Thus $k^2 = 3$, so $k = \sqrt{3}$. Therefore, in 7 hours, it grows by a factor of $(\sqrt{3})^7 = 3^{7/2} = 27\sqrt{3}$.

   (d) Using the results of (a), (b), and (c), the culture grows by a factor of $3^{x/2}$ in x hours.

5. Using the law $(b^x)^y = b^{xy}$, we have $(2^{\sqrt{2}})^{\sqrt{2}} = 2^{(\sqrt{2} \cdot \sqrt{2})} = 2^2 = 4$.

9. Use the laws of exponents to get $5^{\pi/2} \cdot 10^\pi / 15^{-\pi} = 5^{\pi/2} \cdot (5 \cdot 2)^\pi \cdot (5 \cdot 3)^\pi = 5^{5\pi/2} \cdot 2^\pi \cdot 3^\pi = 5^{5\pi/2} \cdot 6^\pi$.

13. Using the laws of exponents, $9^{1/\sqrt{3}} = (3^2)^{1/\sqrt{3}} = 3^{2/\sqrt{3}} = 3^{\sqrt{4/3}}$. Since the base, 3 , is greater than 1 and $\sqrt{2} > \sqrt{4/3}$, $3^{\sqrt{2}}$ is larger.

17.

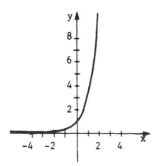

Plot a few points and connect them with a smooth curve.

x	-2	-3/2	-1	-1/2	0	1/2	1	3/2	2
3^x	1/9	0.192	1/3	0.577	1	1.732	3	5.196	9

21.

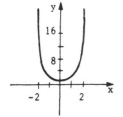

Note that $f(-x) = f(x)$, so we need only plot the graph for $x \geqslant 0$ and then reflect it.

x	0	1/2	1	3/2	7/4	2
$\exp_2 x^2$	1	1.189	2	4.757	8.354	16

25. $\exp_{1/3} x = (1/3)^x = 3^{-x} = \exp_3(-x)$, so the graph of $y = \exp_{1/3} x$ is obtained by reflecting $y = \exp_3 x$ in the y-axis.

29.   Functions (A) and (C) are both strictly negative. For x large and positive, $y = -3^{-x}$ approaches 0 , so (A) matches (b) and (C) matches (c) . Similarly, (B) and (D) are both strictly positive and (B) approaches 0 as x approaches $+\infty$, so (B) matches (d) and (D) matches (a) .

33.

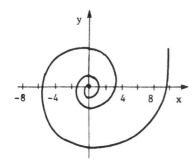

The graph is the same as in Example 7 except that every turn of the spiral is $(1.2)^{2\pi} \approx 3.14$ times as big as the previous one.

37.

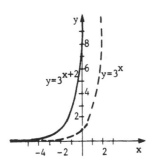

The graph of 3^x is shown in Exercise 17. The graph of $y = 3^{x+2}$ is the same as the graph of $y = 9(3^x)$. This is reasonable since $3^{x+2} = (3^x)(3^2) = 9(3^x)$. In general, shifting the graph of 3^x by k units to the left is the same as graphing $y = 3^{x+k} = (3^k)(3^x)$. Thus, stretching the graph by a factor of 3^k in the y-direction results in the graph of $y = 3^{x+k}$ also.

41. By shifting the graph of 3^x two units to the left, we get the graph of $3^{x+2} = 9(3^x)$. Thus, the area under 3^x between $x = 2$ and $x = 4$ is the same as the area under 3^{x+2} between $x = 0$ and $x = 2$. Therefore, the ratio is $1/9$.

45. Since $b^x > 0$ for all x and all $b > 0$, there is no solution for $2^x = 0$.

SECTION QUIZ

1. Simplify the following expressions:

(a) $[(\exp_2 \pi)(\exp_3 \pi)]^{-1/\pi}$

(b) $5^{\sqrt{3}} \exp_5 \sqrt{9}/25^{\sqrt{3}}$

(c) $y^x x^2 y^x / \sqrt[3]{y^{2x}} \sqrt{x^3}$

2. For each of the following functions, tell (i) whether it is increasing
    or decreasing, (ii) where the y-intercept, if any, is located, (iii)
    whether it has an x-intercept, and (iv) whether it is concave upward
    or downward.

    (a) $(3/2)^{-x} + 5$

    (b) $(3/2)^{x} - 3$

    (c) $(2/3)^{x} - 3$

    (d) $-(3/2)^{-x} + 5$

    (e) $-(2/3)^{x} - 3$

3. For your tenth wedding anniversary, your six-year-old son presents you
    with an ant hill enclosed in glass. You observe that the population
    doubles in about 2 months. Initially, there were 1000 ants.

    (a) Find a function which describes the population after x months.

    (b) Graph the function in part (a).

         26.5 months later, a neighbor gives you a great recipe for
         chocolate ants, but in your anxiety to try out the recipe, you
         break the glass enclosure.

    (c) How many ants are there running around your kitchen and up your
         legs after the accident?

ANSWERS TO PREREQUISITE QUIZ

1. (a) 36

    (b) 64

    (c) 2

ANSWERS TO SECTION QUIZ

1. (a) 1/6

    (b) $5^{3-\sqrt{3}}$

    (c) $y^{4x/3}\sqrt{x}$

2. (a) (i) decreasing; (ii) (0,6) ; (iii) no; (iv) concave upward

    (b) (i) increasing; (ii) (0,-2) ; (iii) yes; (iv) concave upward

    (c) (i) decreasing; (ii) (0,-2) ; (iii) yes; (iv) concave upward

    (d) (i) increasing; (ii) (0,4) ; (iii) yes; (iv) concave downward

    (e) (i) increasing; (ii) (0,-4) ; (iii) no; (iv) concave downward

3. (a) $(1000)2^{x/2}$

    (b)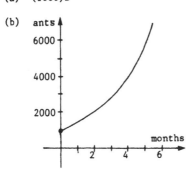

    (c) 9,741,985 ants

6.2 Logarithms

PREREQUISITES

1.   Recall the concept of an inverse function (Section 5.3).

2.   Recall how to sketch exponential functions (Section 6.1).

PREREQUISITE QUIZ

1.   Explain the inverse function test.

2.   (a) Sketch the graph of $\exp_{1/2}x$.

    (b) Sketch the graph of the inverse function of $\exp_{1/2}x$.

    (c) How is the sketch in part (a) related to the graph of $(1/2)^x$?

GOALS

1.   Be able to define logarithms.

2.   Be able to recognize the graphs of and manipulate logarithmic functions.

STUDY HINTS

1.   <u>Definition</u>. The logarithm is simply the exponent. It is defined so
     that if $b^x = y$, then $\log_b y = x$.

2.   <u>Logarithmic graphs</u>. Since the logarithm is an inverse function, you
     can easily recognize the graph of a logarithmic function as a "flipped"
     exponential graph.

3.   <u>Laws of logarithms</u>. It is useful to memorize the three laws. Be care-
     ful <u>not</u> to be tempted to equate $\log_b(x + y)$ with $\log_b x \cdot \log_b y$ or
     with $\log_b x + \log_b y$.

SOLUTIONS TO EVERY OTHER ODD EXERCISE

1. By definition, $\log_2 4 = x$ implies $2^x = 4$; therefore $x = 2$.

5. By definition, $\log_{10}(0.001) = x$ implies $10^x = 0.001$; therefore,

    $x = -3$.

9. By definition, $\log_{1/2} 2 = x$ implies $(1/2)^x = 2$; therefore, $x = -1$.

13.
    The graph of $y = \log_2 x$ is shown in Fig. 6.2.2 . Stretch
    the graph by a factor of 8 in the y-direction.

17. $\log_2(2^8/8^2) = \log_2(2^8/(2^3)^2) = \log_2(2^8/2^6) = \log_2(2^2)$. The logarithm

    is the exponent, which is 2 in this case.

21. By definition, $\log_2(2^b) = x$ implies $2^x = 2^b$; therefore, $x = b$.

25. Try to write 7.5 in terms of 2 , 3 , and 5 . Then apply the laws

    of logarithms. $7.5 = 3 \cdot 5/2$, so $\log_7(7.5) = \log_7(3 \cdot 5/2) = \log_7 3 +$

    $\log_7 5 - \log_7 2 \approx 0.565 + 0.827 - 0.356 = 1.036$.

29. If $\log_b 10 = 2.5$, then $\exp_b(\log_b 10) = \exp_b(2.5)$ or $10 = b^{2.5} =$

    $b^{5/2}$. Raise both sides to the 2/5 power to get $10^{2/5} = (b^{5/2})^{2/5} =$

    b . Now, take the logarithms of both sides to get $2/5 = \log_{10} b$.

    Using a table, we find $\log_{10} 2.51 \approx 0.399674$ and $\log_{10} 2.52 \approx 0.401401$;

    therefore, $b \approx 2.51$.

33. If we let $y = \log_{a^n} x$, then $\exp_{a^n} y = \exp_{a^n}(\log_{a^n} x)$ or $(a^n)^y = a^{ny} =$

    x. Taking the logarithms in base a , we get $ny = \log_a x$ or $y =$

    $(1/n)\log_a x$. Originally, $y = \log_{a^n} x$, so $\log_{a^n} x = (1/n)\log_a x$. Due

    to the definition of the logarithm, we must restrict a to be positive and

    unequal to 1 .

37. By using the laws of logarithms, $2 \log_b (A\sqrt{1 + B}/C^{1/3}B) - \log_b [(B + 1)/AC] =$

$2 [\log_b A + \log_b (1 + B)^{1/2} - \log_b C^{1/3} - \log_b B] - [\log_b (B + 1) - \log_b A - \log_b C] =$

$2 [\log_b A + (1/2)\log_b (1 + B) - (1/3)\log_b C - \log_b B] - [\log_b (B + 1) - \log_b A -$

$\log_b C] = 3 \log_b A - 2 \log_b B + (1/3)\log_b C$.

41. $\log_3 x = 2$ implies that $x = 3^2 = 9$.

45. The logarithm is defined only if $x > 0$ and $1 - x > 0$; therefore, we

must have $0 < x < 1$. From $\log_x (1 - x) = 2$, we get $\exp_x (\log_x (1 - x)) =$

$\exp_x 2$ or $1 - x = x^2$. Then $x^2 + x - 1 = 0$ can be solved with the

quadratic formula, yielding $x = (-1 \pm \sqrt{1 + 4})/2$. Since x must be in

$(0,1)$, $x = (\sqrt{5} - 1)/2$.

49. Successively substitute 2 , 4 , 8 , 10 , 100 , and 1000 for (I_0/I)

in $D = \log_{10}(I_0/I)$ to get $D_2 \approx 0.301$; $D_4 \approx 0.602$; $D_8 \approx 0.903$;

$D_{10} = 1$; $D_{100} = 2$; and $D_{1000} = 3$.

53. We want to find b such that $\log_b 3 = 1/3$, i.e., $b^{1/3} = 3$. Raising

both sides to the third power, we get $b = 3^3 = 27$.

57. Let $y = \log_2 (x - 1)$ and solve for x . Exponentiation gives $2^y = $

$x - 1$, so $x = 2^y + 1$. Changing variables gives the inverse function,

$g(x) = 2^x + 1$. The domain of $g(x)$ is $(-\infty, \infty)$ since 2^x is defined

for all x . Its range is $(1,\infty)$ since $2^x > 0$ for all x .

SECTION QUIZ

1. Determine which statements are true.

(a) $\ln(x + y) = \ln x + \ln y$

(b) $\ln\sqrt{x} = (1/2) \ln x$

(c) $(\ln x)^{1/2} = (1/2) \ln x$

(d) $\log_b xy = \log_b x + \log_b y$

1. (e) $\ln(x/y) = \ln x - \ln y$

    (f) $\log_b(x + y) = \log_b x \cdot \log_b y$

    (g) $\log_b cx = c \log_b x$,   c   constant

2. If $\log_b y = x$, then is $y = b^x$ or is $x = b^y$?

3. If $(-4)^3 = -64$, what is $\log_{(-4)}(-64)$?

4. Your idiotic cousin, Irving, notices that you are studying about loga-
    rithms. He sees the notation "log" and hears you pronouncing "ln" as
    "lawn", so he concludes, "must be reading about building houses." A
    friend tells him that people use the formula $A = P[1 + (r/100n)]^{nt}$ in
    the housing loan business. (This formula is explained on p. 331 of the
    text.)

    (a) Irving wants to borrow money to build a home and asks you how much
         the wood costs. You assume he wants to know $\log_p A$. Write an
         expression for it.

    (b) Next he wants to know how much the grass costs. Compute $\ln A$ for
         him. Sketch the graph of $\ln A$ versus t if P , r , and n
         are positive constants.

    (c) What is $\log_B A$ if $B = 1 + r/100n$?

ANSWERS TO PREREQUISITE QUIZ

1. An inverse function exists if the graph of the function is strictly
    increasing or strictly decreasing.

2.   (a)

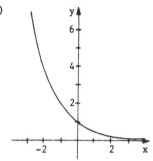

   (b)

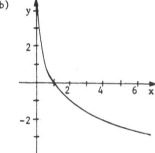

   (c)   They are the same since $\exp_{1/2}x = (1/2)^x$.

ANSWERS TO SECTION QUIZ

1.   b , d , e

2. $y = b^x$

3.   3

4.   (a) $\log_p A = 1 + nt \, \log_p(1 + r/100n)$

4.   (b)   ln A $\ln A = \ln P + nt \ln(1 + r/100n)$

slope $= n \ln(1+ \frac{r}{100n})$

ln P

t

(c) $\log_B A = \log_B P + nt$

6.3 Differentiation of the Exponential and Logarithm Functions

PREREQUISITES

1. Recall the definition of the derivative as a limit (Section 1.3).

2. Recall how to use the chain rule (Section 2.2).

3. Recall how to compute the derivative of an inverse function (Section 5.3).

PREREQUISITE QUIZ

1. Suppose the derivative of $y = f(x)$ at $x = 4$ is $dy/dx = 3$, what is
    the derivative of $f^{-1}(x)$ at the point $f(4)$?

2. State the definition of the derivative as a limit.

3. Write a formula for the derivative of $f(g(x))$ in terms of $f(x)$, $g(x)$,
    $f'(g(x))$, $f(g(x))$, $g'(x)$, and $f'(x)$.

GOALS

1. Be able to differentiate exponential and logarithmic functions.

2. Be able to use the technique of logarithmic differentiation to differ-
    entiate certain functions.

STUDY HINTS

1. <u>The number e</u> . Similar to π , e has a specific value; it is not a
    variable. Its value is approximately 2.718... .

2. <u>Notation and definitions</u>. Logarithms to the base e are called
    natural logarithms; it is usually denoted by $\ln x$ rather than $\log_e x$.
    $\log x$, without a base written, is understood to be the common logarithm,
    which uses 10 as its base.* exp x , without an associated base, is
    understood to be e^x , whereas $\exp_b x$ means b^x .

---

* Except in more advanced courses, where log x means ln x, since $\log_{10} x$ is
  rarely used.

3. <u>Derivatives of exponential functions</u>. $d(e^x)/dx = e^x$ is something that should be memorized. We do not commonly differentiate b^x , so its derivative is easy to forget. Therefore, if we remember that $b = \exp(\ln b)$, then $b^x = \exp(x \ln b)$ and the chain rule may be used to get $d(b^x)/dx = b^x \ln b$. WARNING: n^x and x^n , for constant n , are differentiated differently.

4. <u>Derivatives of logarithmic functions</u>. Memorize $(d/dx)(\ln x) = 1/x$. As with exponentials, logarithms in bases other than e are not often used. One can derive $(d/dx)\log_b x = 1/(\ln b)x$ by writing $b^y = x$ as $e^{(\ln b)y} = x$, so $\ln x = (\ln b)y = (\ln b)\log_b x$, or $\log_b x = \ln x / \ln b$. Differentiating yields $(d/dx)\log_b x = 1/x \ln b$.

5. <u>Logarithmic differentiation</u>. This technique is used mainly to differentiate x to some variable power or complex expressions involving numerical powers. The technique is to take the logarithm of both sides, differentiate, and then solve for dy/dx . Note that $(d/dx)(\ln y) = (1/y)(dy/dx)$. Don't forget to express your answer in terms of the original variable. Study Examples 7 and 8.

6. <u>Integration</u>. As usual, once you have learned the differentiation formulas, you can easily recover the antidifferentiation formulas. Note that we do not have a special antidifferentiation formula to recover $\log_b x$. This is because if we integrate $(d/dx)\log_b x = 1/(\ln b)x$, we get $[1/(\ln b)]\int(dx/x)$, which is one of the given formulas. However, it is still correct to write $\int[dx/(\ln b)x] = \log_b x + C$. Also, note that $\int(dx/x) = \ln |x| + C$, not just $\ln x + C$, unless x is restricted to be positive.

SOLUTIONS TO EVERY OTHER ODD EXERCISE

1. As in Example 1, use $\exp_b'(x) = \exp_b'(0) \exp_b(x)$. Here, $f'(3) =$
    $f'(0)f(3) = f'(0) = 8f'(0)$, so f is increasing 8 times as fast
    at $x = 3$ than at $x = 0$.

5. Use the law that $\ln(e^x) = x$ to get $\ln(e^{x+1}) + \ln(e^2) = (x + 1) + (2) =$
    $x + 3$.

9. Use the law that $\ln(e^x) = x$ to get $e^{4x}[\ln(e^{3x-1}) - \ln(e^{1-x})] =$
    $e^{4x}[(3x - 1) - (1 - x)] = e^{4x}(4x - 2)$.

13. Using the chain rule and the fact that $(d/dx)e^x = e^x$, we get
    $(d/dx)(e^{1-x^2} + x^3) = \exp(1 - x^2) \cdot (d/dx)(1 - x^2) + 3x^2 = -2x \exp(1 - x^2) +$
    $3x^2$.

17. Use the fact that $(d/dx)b^x = (\ln b)b^x$ to get $(d/dx)(3^x - 2^{x-1}) =$
    $(\ln 3)3^x - (\ln 2)2^{x-1}(d/dx)(x - 1) = (\ln 3)3^x - (\ln 2)2^{x-1}$.

21. Use the fact that $(d/dx)\ln x = 1/x$ and the quotient rule to get
    $(d/dx)(\ln x/x) = [(1/x)x - \ln x(1)]/x^2 = (1 - \ln x)/x^2$.

25. Use the chain rule and the fact that $(d/dx)\ln x = 1/x$ to get
    $(d/dx)\ln(2x + 1) = [1/(2x + 1)](d/dx)(2x + 1) = 2/(2x + 1)$.

29. Use the chain rule, quotient rule, and the fact that $(d/dx)\ln x =$
    $1/x$ to get $[(1/\tan 3x)(d/dx)(\tan 3x)(1 + \ln x^2) - (\ln(\tan 3x))(1/x^2) \times$
    $(d/dx)(x^2)]/(1 + \ln x^2)^2 = [3 \sec^2 3x(1 + \ln x^2)/\tan 3x - (2/x)\ln(\tan 3x)]/$
    $(1 + \ln x^2)^2 = [3x \sec^2 3x(1 + \ln x^2) - 2(\tan 3x)\ln(\tan 3x)]/[x \tan 3x \times$
    $(1 + \ln x^2)^2]$.

33. Taking logarithms yields $\ln y = x \ln(\sin x)$. Using the chain rule,
    differentiation gives $(dy/dx)/y = \ln(\sin x) + (x/\sin x)(d/dx)(\sin x) =$
    $\ln(\sin x) + x \cot x$. Hence, $dy/dx = y(\ln(\sin x) + x \cot x) =$
    $(\sin x)^x[\ln(\sin x) + x \cot x]$.

37. Taking logarithms yields $\ln y = (2/3)\ln(x - 2) + (8/7)\ln(4x + 3)$.

    Using the chain rule, differentiation gives $(dy/dx)/y = (2/3)/(x - 2) + (8/7)4/(4x + 3) = 2/3(x - 2) + 32/7(4x + 3) = 2/(3x - 6) + 32/(28x + 21)$.

    Hence, $dy/dx = y[2/(3x - 6) + 32/(28x + 21)] = (x - 2)^{2/3}(4x + 3)^{8/7} \times [2/(3x - 6) + 32/(28x + 21)]$.

41. By the chain rule and product rule, $(d/dx)\exp(x \sin x) = (\sin x + x \cos x)\exp(x \sin x)$.

45. By the chain rule and the fact that $(d/dx)b^x = (\ln b)b^x$, we get $(d/dx)14^{x^2-8 \sin x} = (2x - 8 \cos x)(\ln 14)14^{x^2-8 \sin x}$.

49. Let $u = x^{\sin x}$, so $(d/dx)\cos(x^{\sin x}) = (d/du)\cos u \cdot (du/dx)$. Now use logarithmic differentiation to get du/dx . Begin with $\ln u = \sin x(\ln x)$, so $(du/dx)/u = \cos x (\ln x) + \sin x/x$. Thus, $du/dx = (x^{\sin x})[\cos x (\ln x) + \sin x/x]$ and $(d/dx)\cos(x^{\sin x}) = -[\sin(x^{\sin x})](x^{\sin x})[\cos x (\ln x) + \sin x/x]$.

53. Use the fact that $\exp(\ln x) = x$ to get $(1/x)^{\tan x^2} = x^{-\tan x^2} = e^{-\ln x(\tan x^2)}$. Thus, the derivative is $[-(1/x)\tan x^2 + (-\ln x) \times (2x \sec^2 x^2)]\exp(-\ln x \cdot \tan x^2) = -(1/x)^{\tan x^2}(\tan x^2/x + 2x(\ln x)\sec^2 x^2)$.

57. Use the fact that $\exp(\ln x) = x$, so $3x^{\sqrt{x}} = 3 \exp(\sqrt{x} \ln x)$. Thus, the derivative is $3 \exp(\sqrt{x} \ln x)[(1/2\sqrt{x})\ln x + \sqrt{x}/x] = 3x^{\sqrt{x}}(\ln x/2\sqrt{x} + 1/\sqrt{x})$.

61. Using the fact that $\exp(\ln x) = x$, we have $(\cos x)^x = \exp(x \ln(\cos x))$ so $d(\cos x)^x/dx = \exp(x \ln(\cos x)) \cdot (\ln(\cos x) - x \tan x) = (\cos x)^x(\ln(\cos x) - x \tan x)$. Now, let $y = (\sin x)^{[(\cos x)^x]}$, so by logarithmic differentiation, $\ln y = (\cos x)^x \ln(\sin x)$, and therefore, $(dy/dx)/y = (\cos x)^x(\ln(\cos x) - x \tan x)\ln(\sin x) + (\cos x)^x \cot x$. Hence, the derivative is $dy/dx = (\sin x)^{[(\cos x)^x]}(\cos x)^x \times [(\ln(\cos x) - x \tan x)\ln(\sin x) + \cot x]$.

65. From Example 9(a) , we have $\int e^{ax}dx = (1/a)e^{ax} + C$. Thus, $\int (\cos x +$ $e^{4x})dx = \sin x + e^{4x}/4 + C$.

69. Use division to get $\int [(x^2 + 1)/2x]\,dx = (1/2)\int (x + 1/x)dx = (1/2)(x^2/2 +$ $\ln|x|) + C = x^2/4 + (1/2) \ln |x| + C$.

73. Using the formula $\int b^x dx = b^x/\ln b + C$, we get $\int 3^x dx = 3^x/\ln 3 + C$.

77. $\int_0^1 (x^2 + 3e^x)dx = (x^3/3 + 3e^x)\big|_0^1 = 1/3 + 3e - 3 = 3e - 8/3$.

81. Use the formula $\int b^x dx = b^x/\ln b + C$ to get $\int_0^1 2^x dx = (2^x/\ln 2)\big|_0^1 =$ $(2^1 - 2^0)/\ln 2 = 1/\ln 2$.

85. (a) $(d/dx)x \ln x = \ln x + 1$.

     (b) From part (a), $\int (\ln x + 1)dx = \int \ln x\, dx + \int dx = \int \ln x\, dx + x =$ $x \ln x + C$. Rearrangement yields $\int \ln x\, dx = x \ln x - x + C$.
     Note that the absolute value was not necessary since we were integrating $\ln x$, which we assume exists, so $x > 0$ anyway.

89. According to the fundamental theorem of calculus, the derivative of the integral must equal the integrand.

     (a) $(d/dx)[\ln(x + \sqrt{1 + x^2}) + C] = [1/(x + \sqrt{1 + x^2})][1 + (1/2)(1 +$ $x^2)^{-1/2}(2x)] = [1/(x + \sqrt{1 + x^2})][1 + x/\sqrt{1 + x^2}] = [1/(x + \sqrt{1 + x^2})] \times$ $[(\sqrt{1 + x^2} + x)/\sqrt{1 + x^2}] = 1/\sqrt{1 + x^2}$, which is the integrand.

     (b) $(d/dx)[-\ln|(1 + \sqrt{1 - x^2})/x| + C] = -[x/(1 + \sqrt{1 - x^2})][(1/2)(1 -$ $x^2)^{-1/2}(-2x)x - (1 + \sqrt{1 - x^2})]/x^2 = -[x/(1 + \sqrt{1 - x^2})][-x^2 -$ $\sqrt{1 - x^2} - (1 - x^2)]/x^2\sqrt{1 - x^2} = -[x/(1 + \sqrt{1 - x^2})][(-1)(1 +$ $\sqrt{1 - x^2})]/x^2\sqrt{1 - x^2} = 1/x\sqrt{1 - x^2}$, which is the integrand.

93. We choose $\Delta x = 0.0001$ and denote $f'(\Delta x) = (b^{\Delta x} - 1)/\Delta x$. We know

that $f'(\Delta x) = 0.69$ for $b = 2$ and $f'(\Delta x) = 1.10$ for $b = 3$ and

we wish to find b so that $f'(\Delta x) = 1.00$. Using the method of bi-

section (see Section 3.1), we get the following:

b	2.5	2.75	2.625	2.69	2.70	2.71	2.72	2.73
$f'(\Delta x)$	0.916	1.012	0.965	0.990	0.993	0.997	1.001	1.004

Thus, $e \approx 2.72$.

97. (a) From Exercise 85, $\int_{\epsilon}^{2} \ln x \, dx = (x \ln x - x)\Big|_{\epsilon}^{2}$. For $\epsilon = 1$, 0.1 ,

and 0.01 , the integral is approximately 0.38629, -0.28345 ,

and -0.55765 , respectively.

(b) Define $\int_{0}^{2} \ln x \, dx$ by $\lim_{\epsilon \to 0} \int_{\epsilon}^{2} \ln x \, dx = \lim_{\epsilon \to 0} (x \ln x - x)\Big|_{\epsilon}^{2} =$

$2 \ln 2 - 2 - \lim_{\epsilon \to 0}(\epsilon \ln \epsilon - \epsilon) = 2 \ln 2 - 2 - \lim_{\epsilon \to 0} \epsilon \ln \epsilon$. Numerical

evaluation of $\lim_{x \to 0} x \ln x$ yields the following table:

x	1	0.1	0.01	0.001	0.0001	0.00001	1×10^{-1}
$x \ln x$	0	-0.23	-0.046	-0.0069	-0.00092	-0.00011	-2.3×10^{-9}

Thus, we conclude that $\lim_{x \to 0} x \ln x = 0$ and $\int_{0}^{2} \ln x \, dx = 2 \ln 2 - 2$.

(c) The integral doesn't exist in the ordinary sense because $\ln x$ approach

a negatively infinite value as x approaches 0. Thus, $\ln x$ has no

lower sums on $[0,2]$.

101. $\ln y = n_1 \ln [f_1(x)] + n_2 \ln [f_2(x)] + \ldots + n_k \ln [f_k(x)]$. Differen-

tiation yields $(dy/dx)/y = n_1 f_1'(x)/f_1(x) + n_2 f_2'(x)/f_2(x) + \ldots +$

$n_k f_k'(x)/f_k(x)$, so $dy/dx = y \sum_{i=1}^{k} [n_i f_i'(x)/f_i(x)]$.

SECTION QUIZ

1. Differentiate the following with respect to x :

    (a) x^e

    (b) $\ln e$

    (c) $1/\ln x$

    (d) $3e^2$

    (e) $1/x$

    (f) $1/2^x$

    (g) $(1/2)^x$

    (h) $\log_{10}(3/2)$

    (i) $\log_7 2x$

2. Perform the following integrations:

    (a) $\int e^3 dx$

    (b) $\int x^e dx$

    (c) $\int (3/4x) dx$

    (d) $\int (2/5^x) dx$

3. Find the minimum of $x^{\sqrt[3]{x}}$, $x > 0$.

4. In the midst of a galactic war, an enemy spacecraft was lasered. Your
    computer determines that the disabled ship spirals along the curve given
    by $y = (5x + 3)^3(7x - 8)^4(6x^2 + 12x - 1)^5/(5x^3 - 7x^2 - x + 4)^6$. When
    it reaches $x = 0$,   another laser thrust pushes it out of the galaxy
    along the tangent line. What is the tangent line?

ANSWERS TO PREREQUISITE QUIZ

1. 1/3

2. $f'(x_0) = \lim\limits_{\Delta x \to 0}\{ [f(x_0 + \Delta x) - f(x_0)]/\Delta x\}$

3. $f'(g(x)) \cdot g'(x)$

ANSWERS TO SECTION QUIZ

1. (a) ex^{e-1}

   (b) 0

   (c) $-1/x(\ln x)^2$

   (d) 0

   (e) $-1/x^2$

   (f) $-\ln 2/2^x$

   (g) $-\ln 2/2^x$

   (h) 0

   (i) $1/x \ln 7$

2. (a) $e^3 x + C$

   (b) $x^{e+1}/(e+1) + C$

   (c) $(3/4)\ln |x| + C$

   (d) $-(2 \ln 5)(1/5)^x + C$

3. $e^{-3/e}$

4. $y = 1539x - 27$

6.4 Graphing and Word Problems

PREREQUISITES

1.   Recall how derivatives are used as aids to graphing (Section 3.4).

PREREQUISITE QUIZ

1.   What conclusions about the graph of $y = f(x)$ can you draw if you
     know that:

     (a) $f''(x) < 0$?

     (b) $f'(x) > 0$?

     (c) $\lim\limits_{x \to \infty} f(x) = a$?

GOALS

1.   Be able to graph functions involving exponentials and logarithms.

2.   Be able to define e as a limit.

3.   Be able to compute actual interest rates from compounded interest rates.

STUDY HINTS

1.   Limiting behavior. You should know the three limits listed in the box
     preceding Example 1. Since you will be learning simpler proofs in
     Chapter 11, you probably will not be held responsible for the methods
     of the proofs. Ask your instructor.

2.   Graphing. The same techniques that were introduced in Chapter 3 are
     used to sketch exponentials and logarithms. Use the limiting behaviors
     discussed above to complete the graphs.

3.   Relative rate changes. This quantity is simply the derivative of
     $\ln f(x)$, i.e., f'/f .

4. <u>e as a limit</u>. If you remember that $e = \lim\limits_{h\to 0}(1 + h)^{1/h}$, then letting

    $h = 1/n$ and $h = -1/n$ yields the other two formulas in the box on p. 330.

5. <u>Compound interest</u>. You may find it easier to derive the formula rather

    than memorizing it. At $r\%$ interest compounded n times annually,

    the percent interest during a single period is r/n . Thus, the growth

    factor is $1 + r/100n$. For n periods (1 year), the growth factor

    is $(1 + r/100n)^n$, and over t years, it is $(1 + r/100n)^{nt}$. The

    growth factor $\exp(rt/100)$ is obtained by using the definition of e

    as a limit. Continuously compounded interest is thus a special case

    of exponential growth.

SOLUTIONS TO EVERY OTHER ODD EXERCISE

1.

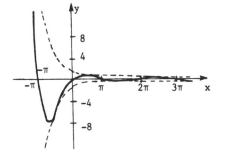

$f'(x) = -e^{-x}\sin x + e^{-x}\cos x = e^{-x}(\cos x - \sin x)$, so the critical points occur where $\cos x = \sin x$ or $\tan x = 1$ or $x = \pi/4 + n\pi$, where n is an integer. The zeros of $f(x)$ occur where $\sin x = 0$ or $x =$ $n\pi$, where n is an integer. Since $-1 \leqslant \sin x \leqslant 1$, $e^{-x}\sin x$ lies between the graphs of e^{-x} and $-e^{-x}$.

5.

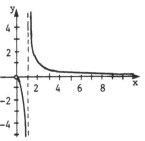

Beginning with $f(x) = y = \log_x 2$, we exponentiate to get $x^y = 2$. Then take the natural logarithm, giving $y \ln x = \ln 2$ or $y = \log_x 2 = \ln 2/\ln x$. The domain is $x > 0$, $x \neq 1$. $f'(x) = -\ln 2/x(\ln x)^2$, so $f(x)$ is always decreasing. $f''(x) =$ $-\ln 2\,(\ln x + 2)/x^2(\ln x)^3$, so the inflection point occurs at $x = 1/e^2$.

9. Using the results of Example 3, $(1/p)(dp/dt) = 0.3 - 0.002t$.

Substituting $t = 2$ corresponding to January 1, 1982, we get

$(1/p)(dp/dt) = 0.3 - (0.002)(2) = 0.296$. Therefore, on January 1,

1982, the company's profits are increasing at a rate of 29.6% per

year.

13. Use the formula $e^a = \lim_{n \to \infty}(1 + a/n)^n$. Using $\exp(\ln x) = x$, we have

$3^{\sqrt{2}} = [\exp(\ln 3)]^{\sqrt{2}} = \exp(\sqrt{2} \ln 3)$. Therefore, $3^{\sqrt{2}} = \lim_{n \to \infty}(1 +$

$\sqrt{2} \ln 3/n)^n$ since $a = \sqrt{2} \ln 3$.

17. The annual percentage increase of funds invested at r% per year com-

pounded continuously is $100(e^{r/100} - 1)\%$. Therefore, if $\dot{r} = 8$,

then the actual yield is $100(e^{0.08} - 1) \approx 8.33\%$.

21. The tangent line is $y = y_0 + (dy/dx)\big|_{x_0} (x - x_0)$. $dy/dx = 2xe^{2x} +$

e^{2x} , so $(dy/dx)\big|_1 = 3e^2$. Also, at $x = 1$, $y = e^2$, so the

tangent line is $y = e^2 + (3e^2)(x - 1) = 3e^2 x - 2e^2$.

25. The tangent line at (x_0, y_0) is $y = y_0 + (dy/dx)\big|_{x_0} (x - x_0)$. $dy/dx =$

$2x/(x^2 + 1)$, so $(dy/dx)\big|_1 = 1$. Also, at $x = 1$, $y = \ln 2$, so

the tangent line is $y = \ln 2 + 1(x - 1) = x + (\ln 2 - 1)$.

29. Using $\exp(\ln x) = x$, we get $y = x^x = \exp(x \ln x)$, so $y' =$

$(\ln x + 1)\exp(x \ln x)$. This vanishes when $x = 1/e$. At $x = 1/e$,

$y = (1/e)^{1/e} = e^{-1/e} \approx 0.692$. Since this is the only critical point

and the limits of x^x at 0 and ∞ are larger, 0.692 is the minimum

value.

33. Apply the hint by first finding $dp/dx = -2116(0.0000318)\exp(-0.0000318x)$,

which is approximately $-0.0672\exp(-0.0636) \approx -0.0631$ when $x = 2000$.

At the time in question, $dx/dt = 10$, so $dp/dt = (dp/dx)(dx/dt) \approx$

$-(0.0631)(10) = -0.631$.

37. (a) $g'(x) = f(1 + 1/x) + xf'(1 + 1/x)(-1/x^2) = f(1 + 1/x) -$

   $f'(1 + 1/x)/x$. The mean value theorem states that there exists

   some x_0 in $(1, 1 + 1/x)$ at which $f'(x_0) = [f(1 + 1/x) -$

   $f(1)]/(1/x)$. Since $f(1) = 0$, $f'(x_0)/x = f(1 + 1/x)$. Sub-

   stitute this into $g'(x) = [f'(x_0) - f'(1 + 1/x)]/x$. Since $f'(x)$

   is a decreasing function and $x_0 < 1 + 1/x$, $f'(x_0) - f'(1 + 1/x) >$

   0 . If $x \geqslant 1$, $g'(x) > 0$, so $g(x)$ is increasing on $[1, \infty)$.

   (b) $g(x) = x \ln(1 + 1/x)$ is increasing on $[1, \infty)$.

   (c) Let $x = n/a$. Then $g(x) = (n/a) \ln(1 + a/n) = (1/a)\ln(1 + a/n)^n$

   is increasing on $[1, \infty)$. To show that $(1 + a/n)^n$ is increasing,

   multiply by a and exponentiate, giving $e^{ag(x)} = (1 + a/n)^n$.

   Since $a > 0$ and $g(x)$ is increasing, $e^{ag(x)}$ and therefore

   $(1 + a/n)^n$ are also increasing.

SECTION QUIZ

1. Sketch the graph of $y = x \ln x^2$.

2. Sketch the graph of $y = \sqrt[3]{x} \, e^x$.

3. Express e^a as a limit.

4. Having eyed a $2 million mansion, you decide that you must have it in

    ten years. Long-term interest rates will pay 12% annually with

    continuous compounding. How much must be put into the savings account

    to get $2 million in ten years?

ANSWERS TO PREREQUISITE QUIZ

1.   (a) f(x) is concave downward.

     (b) f(x) is increasing.

     (c) y = a is a horizontal asymptote.

ANSWERS TO SECTION QUIZ

1.

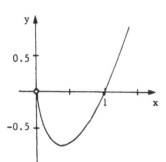

2.

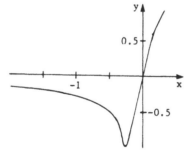

3. $\lim_{h\to 0}(1 + h)^{a/h} = \lim_{n\to\infty}(1 + 1/n)^{an} = \lim_{n\to\infty}(1 - 1/n)^{-an}$

4. $602,388.43

6.R Review Exercises for Chapter 6

SOLUTIONS TO EVERY OTHER ODD EXERCISE

1. Using $a^{x+y} = a^x a^y$ and $(a + b)(a - b) = a^2 - b^2$, we get
    $(x^\pi + x^{-\pi})(x^\pi - x^{-\pi}) = x^{2\pi} - x^{-2\pi}$.

5. Using the fact that $\ln(e^x) = x$, we get $\ln(e^3) + (1/2)\ln(e^{-5}) = 3 + (1/2)(-5) = 1/2$.

9. Use the chain rule and $(d/dx)e^x = e^x$ to get $(d/dx)\exp(x^3) = 3x^2\exp(x^3)$.

13. Use the chain rule and $(d/dx)e^x = e^x$ to get $(d/dx)\exp(\cos 2x) = (-2 \sin 2x)\exp(\cos 2x)$.

17. Use the chain rule and $(d/dx)e^x = e^x$ to get $(d/dx)\exp(6x) = 6\exp(6x)$.

21. Use the quotient rule to get $(d/dx)[\sin(e^x)/(e^x + x^2)] = [((e^x) \cos (e^x))(e^x + x^2) - (\sin (e^x))(e^x + 2x)]/(e^x + x^2)^2$.

25. Use the chain rule and $(d/dx)e^x = e^x$ to get $(d/dx)\exp(\cos x + x) = (-\sin x + 1)\exp(\cos x + x)$.

29. Use the chain rule, product rule, and $(d/dx) \ln x = 1/x$ to get $(d/dx)x \ln(x + 3) = \ln(x + 3) + x/(x + 3)$.

33. Exponentiate to get $5x = 3^y = \exp (y \ln 3)$. Take natural logarithms to get $\ln 5 + \ln x = y \ln 3$. Differentiate to get $1/x = \ln 3(dy/dx)$, i.e., $dy/dx = 1/x \ln 3$.

37. Use the reciprocal rule to get $-[2(\ln t)/t]/[(\ln t)^2 + 3]^2 = -2 \ln t/ t[(\ln t)^2 + 3]^2$.

41. Use the formula $\int (dx/x) = \ln |x| + C$ to get $\int(\cos x + 1/3x)dx = \sin x + (1/3)\ln |x| + C$.

45. Use division to get $\int_1^2 [(x + x^2 \sin \pi x + 1)/x^2] dx = \int_1^2 (x^{-1} + \sin \pi x + x^{-2}) dx = (\ln |x| - \cos \pi x/\pi - 1/x)|_1^2 = \ln 2 - 2/\pi + 1/2$. $\sin \pi x$ was integrated by guessing it was a $\cos \pi x$ and differentiation was used to find a .

49. Let $y = (\ln x)^x$, so $\ln y = x \ln(\ln x)$. Now apply the chain rule to get $(dy/dx)/y = \ln(\ln x) + (x/\ln x)(1/x) = \ln (\ln x) + 1/\ln x$. Therefore, $dy/dx = (\ln x)^x [\ln(\ln x) + 1/\ln x]$.

53.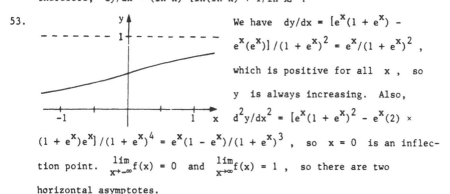

We have $dy/dx = [e^x(1 + e^x) - e^x(e^x)]/(1 + e^x)^2 = e^x/(1 + e^x)^2$, which is positive for all x , so y is always increasing. Also, $d^2y/dx^2 = [e^x(1 + e^x)^2 - e^x(2) \times (1 + e^x)e^x]/(1 + e^x)^4 = e^x(1 - e^x)/(1 + e^x)^3$, so $x = 0$ is an inflection point. $\lim_{x \to -\infty} f(x) = 0$ and $\lim_{x \to +\infty} f(x) = 1$, so there are two horizontal asymptotes.

57. Rearrange to get $e^{xy} = 1 + xy$. Then, differentiate implicitly to get $(y + x(dy/dx))e^{xy} = y + x(dy/dx)$. Rearrange again to get $xe^{xy}(dy/dx) - x(dy/dx) = y - ye^{xy} = (dy/dx)(xe^{xy} - x)$. Therefore, $dy/dx = (y - ye^{xy})/(xe^{xy} - x) = y(1 - e^{xy})/x(e^{xy} - 1) = -y/x$.

61. The tangent line is $y = y_0 + m(x - x_0)$. Here, $m = (dy/dx)|_0 = [(1)\exp(3x^2 + 4x) + (x + 1)(6x + 4)\exp(3x^2 + 4x)]|_0 = 5$. Hence, the tangent line is $y = 5x + 1$.

65. The derivative is $[e^x(x + 1) - e^x(1)]/(x + 1)^2 = xe^x/(x + 1)^2$; therefore, $\int [xe^x/(x + 1)^2] dx = e^x/(x + 1) + C$.

69. Apply the formula $e^a = \lim_{n \to +\infty}(1 + a/n)^n$. Here, $a = 10$, so the limit is e^{10} .

73. With continuous compounding, the interest rate is $100(e^{r/100} - 1)\%$.

   Thus, we get $e^{r/100} - 1 = 0.08$ or $e^{r/100} = 1.08$, i.e., $r = 100 \ln (1.08) \approx 7.70\%$.

77. (a) After one interval, $t = 1$ and $A(t) = A(1) = A_0$. At the end of
   the second interval, $A(2) = A_0 + A_0(1 + i/n)^n$. At the end of
   the third interval, $A(3) = A_0 + A(2)(1 + i/n)^n = A_0 + A_0(1 + i/n)^n + A_0(1 + i/n)^{2n}$. Similarly, $A(t) = A_0 \sum\limits_{k=0}^{t-1} (1 + i/n)^{nk}$. Using the

   hint, $\sum\limits_{k=0}^{t-1} (1 + i/n)^{nk} = ((1 + i/n)^{nt} - 1)/((1 + i/n)^n - 1)$, so

   $A(t) = A_0((1 + i/n)^{nt} - 1)/((1 + i/n)^n - 1)$.

   (b) Here, $A_0 = 400$, $n = 1$, $i = 0.07/4$ and $t = 24$. Hence $A = 400((1 + 7/400)^{24} - 1)/((1 + 7/400) - 1) \approx (400)^2(0.516/7) \approx \$11,804.41$

81. (a) Differentiate $P(t)$: $dP/dt = -a(b + ((a/P_0) - b)e^{-at})^{-2}(a/P_0 - b) \times (-ae^{-at})$. Add $0 = a^2b - a^2b$ to get $dP/dt = [(a^2(a/P_0 - b)e^{-at} + a^2b) - a^2b] \cdot (b + (a/P_0 - b)e^{-at})^{-2}$. Use the distributive law, yieldin

   $dP/dt = a^2((a/P_0 - b)e^{-at} + b)(b + (a/P_0 - b)e^{-at})^{-2} - b(a/(b + (a/P_0 - b)e^{-at}))^2$. Cancel the factors of a^2 and substitute P ,

   giving $dP/dt = a^2/(b + (a/P_0 - b)e^{-at}) - bP^2 = aP - bP^2 = P(a - bP)$.
   When $t = 0$, $P(0) = a/(b + (a/P_0 - b)) = a/(a/P_0) = P_0$.

   (b) $\lim\limits_{t \to \infty} P(t) = a/(b + (a/P_0 - b) \cdot 0) = a/b$.

85. (a) Use mathematical induction. If $n = 1$, the relation is true,
   since $b \geqslant 1 + b - 1 = b$. Suppose the relation holds for some
   n: $b \geqslant 1 + n(b - 1)$. Multiply by b to get $b \cdot b^n = b^{n+1} \geqslant b + nb(b - 1) = 1 + (b - 1) + nb(b - 1) = 1 + (nb + 1)(b - 1)$. Since
   $b > 1$, $nb + 1 > n + 1$, so $b^{n+1} \geqslant 1 + (n + 1)(b - 1)$. Thus if
   the relation holds for some n , it also holds for $n + 1$. Since

85. (a) (continued)

it also holds for $n = 1$, it holds for all n . Take reciprocals

to show that $b^{-n} \leqslant 1/(1 + n(b - 1))$.

(b) $\lim_{x \to \infty} b^x \geqslant \lim_{x \to \infty} [1 + x(b - 1)] = \lim_{x \to \infty} 1 + \lim_{x \to \infty} x(b - 1) = \lim_{x \to \infty} 1 +$

$(b - 1)\lim_{x \to \infty} x = \lim_{x \to \infty} (1 + x) = \lim_{x \to \infty} x = \infty$; therefore, $\lim_{x \to \infty} b^x = \infty$.

We know that $0 < b^{-n} \leqslant 1/[1 + n(b - 1)]$, so $\lim_{x \to -\infty} b^x = \lim_{x \to \infty} b^{-x} \leqslant$

$\lim_{x \to \infty} \{1/[1 + x(b - 1)]\} = 1/\lim_{x \to \infty} [1 + x(b - 1)] = 1/[\lim_{x \to \infty} 1 +$

$(b - 1)\lim_{x \to \infty} x] = 1/\lim_{x \to \infty} (1 + x) = 1/\lim_{x \to \infty} x = \lim_{x \to \infty} (1/x) = 0$; therefore,

$\lim_{x \to -\infty} b^x = 0$.

89. (a) By taking logarithms, we get $x \geqslant n \ln x$ for $x \geqslant (n + 1)!$

Rearrangement yields $x/\ln x \geqslant n$.

(b) Taking the reciprocal of the inequality in part (a) gives $\ln x/x \leqslant$

$1/n$ when $x \geqslant (n + 1)!$ When $n \to \infty$, $x \to \infty$; and both $\ln x$ and

x are positive, so $0 \leqslant \lim_{x \to \infty} (\ln x/x) \leqslant \lim_{n \to \infty} (1/n) = 0$. Consequently,

$\lim_{x \to \infty} (\ln x/x) = 0$.

93. Take the natural logarithm of both sides. We want to show that

$(x - 2) \ln 3 > \ln 2 + 2 \ln x$ or $(x - 2) > \ln 2/\ln 3 +$

$(2/\ln 3) \ln x$ or $x - (2/\ln 3) \ln x > \ln 2/\ln 3 + 2$. Let $f(x) =$

$x - (2/\ln 3) \ln x$, so what we want to prove is $f(x) > 2 +$

$\ln 2/\ln 3 \approx 2.63$ if $x \geqslant 7$. Note that $f'(x) = 1 - (2/\ln 3)(1/x)$,

so $f'(x) > 0$ for $x \geqslant 2/\ln 3 \approx 1.82$. Thus, since f is increasing,

$f(x) > f(7)$ if $x \geqslant 7$. But $f(7) = 7 - (2/\ln 3)\ln 7 \approx 3.46 > 2.63$,

so we get our result. We also find $f(6) = 6 - (2/\ln 3)\ln 6 \approx 2.74 >$

2.63, so in fact the statement holds if $x \geqslant 6$. (Numerically

experimenting shows that it is actually valid if $x \geqslant 5.8452\ldots$.)

TEST FOR CHAPTER 6

1. True or false.

   (a) exp a + exp b = exp(a + b) .

   (b) The domain of ln |x| is only x > 0 .

   (c) exp |x| has symmetry in the y-axis.

   (d) $\int_{-e}^{e} (dx/x) = \ln |x| \,\Big|_{-e}^{e}$.

   (e) $\log_a 1 = 0$ for all real a > 0 .

2. Differentiate the following functions:

   (a) e/x

   (b) (exp x)(log$_{10}$x)

   (c) $\sqrt{5^x}$

   (d) ln (4x)

3. Evaluate the following:

   (a) $\int 8e^{8t} dt$

   (b) $\int [4 \, dx/(x - 5)]$

   (c) $\int_1^{\exp 2} (e/x) dx$

4. Sketch the graph of $y = \ln(x^2 + 1)$.

5. Write an equation of the form $y = \pm A \exp(\pm Bx)$ with the following
   properties, where A > 0 and B > 0 are constants.

   (a) The graph is increasing and concave downward.

   (b) The graph is decreasing and concave downward.

   (c) The graph is increasing and concave upward.

   (d) The graph is decreasing and concave upward.

6. Without using a calculator, approximate the following by using the

   fact that $\ln 2 \approx 0.693$, $\ln 3 \approx 1.099$, and $\ln 5 \approx 1.609$.

   (a) $[(d/dx)\int_4^x t \ln t \, dt]\big|_3$

   (b) $(d/du)(\ln u/u)\big|_{10}$

   (c) $\int_1^5 x(\ln 10)dx$

7. Compute the following:

   (a) $(d/dt)(\exp t + \ln t)$

   (b) $(d/dx)[(4x^3 + 9)^2(8x^5)(2/x + 3)^3/(6)^5(x^3 - x)^6]$

   (c) $(d/dx)[\exp(\ln(5x))]$

   (d) $(d/dy)[e^y(y + 5)^3(y^2 - y)(y/e^{2y})^3]$

8. Multiple choice. More than one may be correct.

   (a) The graph of $y = -\ln(x/2)$ has a _____ asymptote; it is _____ .

       (i)   vertical, $y = 0$

       (ii) horizontal, $y = 0$

       (iii) vertical, $x = 0$

       (iv) horizontal, $x = 0$

   (b) Which is the reflection of $e^{3x} + 3$ across the y-axis?

       (i) $(1/e)^{3x} + 3$

       (ii) $(1/e)^{3x} - 3$

       (iii) $-e^{3x} + 3$

       (iv) $e^{-3x} + 3$

9. If P dollars is invested at r% interest compounded daily, the amount

   after t years is given by $A = P(1 + r/365)^{365t}$.

   (a) Make a graph to show how long it takes to double your money at

       r% interest.

   (b) Suppose 365 is replaced by n in the formula. Simplify the

       formula if $n \rightarrow \infty$.

10. One day, the Gabber went to see his doctor because of a sore tongue.
    It was discovered that the cells had been cancerous due to excess
    gossiping. Also, it was known that the cell population doubled for
    every five pieces of juicy news coming out of the Gabber's mouth.

    (a) Find the growth factor for the cancerous cells in terms of the
         number of pieces of news n .

    (b) There are now 10^9 cells. At what rate are the cells growing?

    (c) 10^{12} cancer cells is lethal. How much more gossiping can the
         Gabber do?

ANSWERS TO CHAPTER TEST

1. (a) False; exp a + exp b = exp ab

    (b) False; the domain includes x < 0 .

    (c) True

    (d) False; ln $|x|$ is not continuous at x = 0 .

    (e) True

2. (a) $-e/x^2$

    (b) $(\exp x)(\log_{10}x + 1/x \ln 10)$

    (c) $(\ln 5)\sqrt{5x}/2$

    (d) $1/x$

3. (a) $e^{8t} + C$

    (b) $4 \ln |x - 5| + C$

    (c) 2e

4.

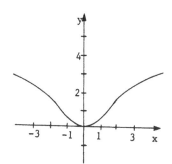

5.   (a) y = -A exp(-Bx)

     (b) y = -A exp (Bx)

     (c) y = A exp (Bx)

     (d) y = A exp (-Bx)

6.   (a) 3.297

     (b) -0.01302

     (c) 27.624

7.   (a) exp t + 1/t

     (b) $[(4x^3 + 9)^2(8x^5)(2/x + 3)^3/(6)^5(x^3 - x)^6] [24x^2/(4x^3 + 9) + 5/x -$
          $(6/x^2)/(2/x + 3) - 6(3x^2 - 1)/(x^3 - x)]$

     (c) 5

     (d) $[e^y(y + 5)^3(y^2 - y)(y/e^{2y})^3] [1 + 3/(y + 5) + (2y - 1)/(y^2 - y) +$
          $3/y - 6]$

8.   (a) iii

     (b) i, iv

9.   (a)

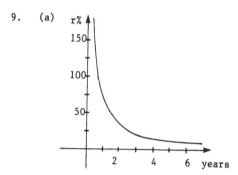

(b) $A = Pe^{rt}$

10. (a) $2^{n/5}$

(b) $(\ln 2/5)10^9$

(c) $5 \ln 1000/\ln 2$ pieces of news.

COMPREHENSIVE TEST FOR CHAPTERS 1-6 (Time limit: 3 hours)

1.   True or false. If false, explain why.

(a)   The area between the x-axis and the curve $x^2 - x$ on $[0,1]$ is
$\int_0^1 (x^2 - x) dx$.

(b) $x^4 - 3x - 2 = 0$ has a solution for x between 1 and 2 .

(c) $(d/dx)(1-x)^3 = 3(1 - x)^2$.

(d)   If f and g are both increasing on $[a,b]$, then $f + g$ is
also increasing on $[a,b]$.

(e)   In general, $\int_a^b f(x) g(x) dx = \int_a^b f(x) dx \cdot \int_a^b g(x) dx$.

(f)   All continuous functions on $[a,b]$ are integrable.

(g)   If $n > 0$ is constant, then $(d/dx)n^x$ is xn^{x-1} .

(h) $y = \sin x$ satisfies $y'' + y = 0$.

(i)   The slope of $r = \sin 2\theta$, graphed in the xy-plane, is
$(2 \cos 2\theta)\big|_{\pi/5}$ at $\theta = \pi/5$.

2.   Fill in the blank.

(a)   The fundamental theorem of calculus states that $\int_a^b f(x) dx =$
_____ , where $F'(x) = f(x)$.

(b)   The point with cartesian coordinates $(-3,3)$ has polar
coordinates _____ .

(c)   Simplify: $\exp(4 \ln(\exp x^2)) \cdot \ln(\exp(\exp(4x^2))) =$
_____ .

(d)   dr/dw is the derivative of _____ with respect
to _____ .

(e)   The point with polar coordinates $(-2,2)$ lies in the
_____ quadrant.

3.   Differentiate the following functions of x .

(a) $(\sin 2x \cos x)^{3/2}$

(b) $e^{3x}/\ln (x + 2)$

(c)   The inverse of $x^5 + x^3 + 1$ at $x = 1$.

4.   Multiple choice.

(a)   The derivative of $\cos^{-1}x$ is:

(i) $-\sin x$

(ii) $-1\sqrt{1 - x^2}$

(iii) $-1/\sqrt{x^2 - 1}$

(iv) $1/\sqrt{1 - x^2}$

(b)

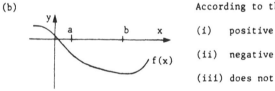

According to the figure, $\int_a^b f(x)dx$ is:

(i)   positive

(ii)   negative

(iii) does not exist

(iv)   unknown; insufficient information

(c)   The antiderivative F of $f(x) = x^5 - x^3 + x - 2$ such that

$F'(x) = f(x)$ and $F(1) = 0$ is:

(i) $x^6/6 - x^4/4 + x^2/2 - 2x + C$

(ii) $5x^4 - 3x^2 + 1 + C$

(iii) $x^6/6 - x^4/4 + x^2/2 - 2x - 19/12$

(iv) $x^6/6 - x^4/4 + x^2/2 - 2x + 19/12$

(d)   If $f'(a) = 0$ and $f''(a) > 0$, then:

(i) $f(a)$ is a local minimum .

(ii) $f(a)$ is a local maximum .

(iii) $f(x)$ is increasing at $x = a$.

(iv)   no conclusion can be made .

4. (e) Suppose $g'(t) > 0$ on $[a,b]$. Then $\int_b^a g(t)dt$ is:

   (i) positive

   (ii) negative

   (iii) zero

   (iv) unknown; need more information

5. (a) Differentiate x^x .

   (b) Evaluate $\int_1^2 [(3x^2 + 2x)/x]\,dx$.

   (c) Differentiate $3^x/\sin^{-1}x$.

   (d) Evaluate $\int [3dy/(1 + y^2)]$.

6. Consider the function $f(x) = (x + 3)/(x - 1)$.

   (a) Discuss its asymptotes .

   (b) Discuss its critical points.

   (c) Where is $f(x)$ increasing? Decreasing?

   (d) Where is $f(x)$ concave upward? Downward?

   (e) Sketch the graph of $f(x)$.

7. Short answer questions.

   (a) Compute $\displaystyle\sum_{i=6}^{101} [(i - 1)^2 - i^2]$.

   (b) Find $(d/dt)\int_t^3 xe^x\cos(x + 2)dx$.

   (c) Approximate $(1.11)^5$ by using the linear approximation.

   (d) Find dy/dx if $(y + 3)x = x^2 y^3 - 5$.

   (e) Sketch the graph of $r = 3\cos\theta$ in the xy-plane.

8. A tightrope walker needs to walk from the top of a 10 m. building to the ground and then back up to the top of a 20 m. building. The bases of the buildings are separated by 50 m. Where should the rope be placed on the ground between the buildings to minimize the distance walked?

9. Integration word problems.

    (a) Find the area of the region bounded by $y = |x| + 2$ and $y = x^2$.

    (b) A millionaire is spending money at the rate of $(e^x/3 + \cos x + 3)$ thousand dollars per hour. How much money does he spend between hours 2 and 3 ?

10. Mount Olympus is located 400 m. above sea level. Sitting on his throne, Zeus spots Mercury running off at sea level at the rate of 50m./min. When Zeus spots Mercury, they are separated by 500 m. How fast must Zeus rotate his head upward (in radians/minute) to keep his eyes on Mercury?

ANSWERS TO COMPREHENSIVE TEST

1. (a) False; $x^2 - x \leqslant 0$ on $[0,1]$, so the area is $-\int_0^1 (x^2 - x)dx$.

    (b) True; use the intermediate value theorem.

    (c) False; the chain rule requires another factor of -1 .

    (d) False; let $f(x) = g(x) = x$ on $[-1,0]$.

    (e) False; let $f(x) = x$ and $g(x) = x$ on $[0,1]$.

    (f) True

    (g) False; $(d/dx)n^x = (\ln n)n^x$.

    (h) True

    (i) False; the slope is $[\tan(\pi/5)(2\cos(2\pi/5)) + \sin(2\pi/5)]/$
         $[2\cos(2\pi/5) - \sin(2\pi/5)\tan(\pi/5)]$.

2. (a) $F(b) - F(a)$

    (b) $(3\sqrt{2}, 3\pi/4)$

    (c) $\exp(8x^2)$

    (d) $r;w$

    (e) fourth

3. (a) $(3/2)(\sin 2x \cos x)^{1/2}(2 \cos 2x \cos x - \sin 2x \sin x)$

   (b) $e^{3x}[3(x + 2)\ln(x + 2) - 1]/(x + 2)[\ln(x + 2)]^2$.

   (c) $1/8$

4. (a) ii

   (b) ii

   (c) iv

   (d) i

   (e) iv

5. (a) $x^x(\ln x + 1)$

   (b) $13/2$

   (c) $[(\ln 3)3^x \sin^{-1}x - 3^x/\sqrt{1 - x^2}]/(\sin^{-1}x)^2$

   (d) $3 \tan^{-1}y + C$

6. (a) Horizontal asymptote: $y = 1$; vertical asymptote: $x = 1$.

   (b) There are no critical points.

   (c) Decreasing on $(-\infty,1)$ and $(1,\infty)$

   (d) Concave upward on $(1,\infty)$; concave downward on $(-\infty,1)$

   (e)

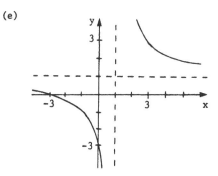

7. (a) -10176

   (b) $-te^t\cos(t + 2)$

   (c) 1.55

7.   (d) $(y + 3 - 2xy^3)/(3x^2y^2 - x)$

     (e)

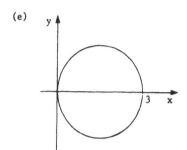

8.   50/3 meters from the 10 m. building.

9.   (a)   20/3

     (b) $[(e^3 - e^2)/3 + \sin(3) - \sin(2) + 3]$ thousand dollars $\approx$ 6464 dollars

10. 0.08 radians/minute.

Printed in the United States
By Bookmasters